Radio Frequency Identification (RFID) Technology and Application in Fashion and Textile Supply Chain

Textile Institute Professional Publications
Series Editor: The Textile Institute

Care and Maintenance of Textile Products Including Apparel and Protective Clothing
Rajkishore Nayak, Saminathan Ratnapandian

For more information about this series, please visit: www.crcpress.com/Textile-Institute-Professional-Publications/book-series/TIPP

Radio Frequency Identification (RFID) Technology and Application in Fashion and Textile Supply Chain

by Rajkishore Nayak

CRC Press
Taylor & Francis Group
Boca Raton London New York

CRC Press is an imprint of the
Taylor & Francis Group, an **informa** business

CRC Press
Taylor & Francis Group
6000 Broken Sound Parkway NW, Suite 300
Boca Raton, FL 33487-2742

Printed on acid-free paper

International Standard Book Number-13978-0-8153-7623-1 (Paperback)

Library of Congress Cataloging-in-Publication Data

Names: Nayak, Rajkishore, author.
Title: Radio frequency identification (RFID) technology and application in fashion and textile supply chain / by Rajkishore Nayak.
Description: Boca Raton, FL : CRC Press/Taylor & Francis, 2019. | Includes bibliographical references and index.
Identifiers: LCCN 2018042968| ISBN 9780815376231 (pbk. : alk. paper) | ISBN 9780815376279 (hardback : alk. paper) | ISBN 9781351238250 (ebook)
Subjects: LCSH: Clothing trade—Inventory control.—Data processing. | Textile industry—Materials management—Data processing. | Radio frequency identification systems. | Business logistics.
Classification: LCC HD9940.A2 N39 2019 | DDC 677.0068/7—dc23
LC record available at https://lccn.loc.gov/2018042968

Visit the Taylor & Francis Web site at
http://www.taylorandfrancis.com

and the CRC Press Web site at
http://www.crcpress.com

Contents

List of abbreviations

1D	One-Dimensional
2D	Two-Dimensional
AAR	Association of American Railroads
AI	Artificial Intelligence
AIDC	Automatic Identification and Data Capture
AMR	Advanced Market Research
AQL	Acceptance Quality Level
ASN	Advanced Shipping Notice
AWEX	Australian Wool Exchange
AWTA	Australian Wool Testing Authority
CASPIAN	Consumers Against Supermarket Privacy Invasion and Numbering
CCD	Charge Coupled Device
CD-ROM	Compact Disk-Read Only Memory
CIE	Commission Internationale De l'Elcairage
CIF	Cost, Insurance and Freight
CPU	Central Processing Unit
CRM	Customer Relationship Management
DC	Distribution Center
CDRW	Compact Disc-Rewritable
DCM	Data Capturing Module
DoD	Department of Defense
DPI	Department of Primary Industries
DSM	Decision Support Module
EDI	Electronic Data Interchange
EEPROM	Electrically Erasable Programmable Read-Only Memory
EFF	Electronic Frontier Foundation
EPC	Electronic Product Code
EPIC	Electronic Privacy Information Center
ERP	Enterprise Resource Planning
EXW	Ex-Works
FDA	Food and Drug Administration
FOB	Free-on-Board

FTC	Federal Trade Commission
GHG	Greenhouse Gas
HF	High Frequency
IC	Integrated Circuit
IED	Industrial Engineering Department
IFF	Identify Friend or Foe
IoT	Internet of Things
ISO	International Standards Organization
IT	Information Technology
JIT	Just-in-Time
LED	Light Emitting Diode
LF	Low Frequency
LSP	Logistics Service Provider
MEG	Monoethylene Glycol
MES	Manufacturing Execution System
MFA	Multi-Fiber Arrangement
MPS	Modular Production Systems
MRO	Maintenance, Repair and Operations
NFC	Near Field Communication
NLIS	National Livestock Identification System
NYPA	New York Port Authority
OTB	Open to Buy
PBS	Progressive Bundle Systems
POS	Point-of-Sale
PPC	Production Planning and Control
PTA	Purified Terephthalic Acid
QC	Quality Control
QR	Quick Response
RAS	Resource Allocation System
RF	Radio Frequency
RFID	Radio Frequency Identification
RN	Registered Identification Number
RO	Read-Only
ROI	Return on Investment
RW	Read-Write
SAM	Standard Allowed Minutes
SKU	Stock Keeping Unit
SME	Small and Medium-Sized Enterprise
TBL	Triple Bottom Line
TPL	Third-Party Logistics
TPU	Thermoplastic Polyurethane
UCC	Uniform Code Council
UHF	Ultra High Frequency
UPS	Unit Production System

UV	Ultraviolet
WIP	Work in Progress
WMS	Warehouse Management System
WORM	Write-Once-Read-Many
WTO	World Trade Organization

Preface

Radio Frequency Identification (RFID) is gaining popularity with many organizations as a solution for effectively managing their supply chain activities. The first use of the RFID technology can be dated back to World War II, when fighter planes used it to uniquely identify enemy planes. The first commercial adopters were Walmart and the US Department of Defense (DoD). Walmart mandated the adoption of RFID technology for all its suppliers in 2003 for easy inventory management, which was followed by the DoD for pallet-level tracking. Since then, Walmart and DoD have realized the benefits of implementing RFID technology.

Witnessing the benefits Walmart and DoD were gaining from RFID technology, other global retailers and manufacturers began adopting the technology to also realize its benefits, which include: accuracy of information, prevention of loss, product visibility, easy traceability, reduced labor cost and reduced time. As a result, RFID technology is now being widely used in many sectors, such as: medicine, civil, electronics, automotive, food, logistics, healthcare, tourism, library, aviation, textile and fashion.

RFID technology is a remote identification system using radio waves of different frequencies. It is considered one of the most influential innovations of the 20th century. RFID technology is based on a wireless communication system, which uses electromagnetic fields for automatic identification of objects, animals or even people. An RFID tag contains electronically stored information, which can be accessed by a reader when it is operated within the range of the tag. It took several years for researchers and manufacturers to develop a functional RFID system that operates similarly to barcode technology. After the development of the functional system, RFID technology is being rapidly adopted by many industries, including fashion and textiles, to manage their production and supply chain activities.

Although many of us use various RFID systems in our daily lives, we may not be aware of the technology or its principles of operation. Some common examples of RFID in daily life include: access cards to offices; train tickets; cards to operate lifts; library cards; and even contactless car keys used to start a car. This technology has the potential to be applied in many production and distribution operations for improved accuracy and speed of information retrieval. The major driving factors for adoption of RFID technology are efficiency in tracing and tracking; reduced time and labor cost; reduced shrinkage; increased visibility; and accuracy of inventory levels.

Although RFID technology was coined during World War II, wider commercial applications gained momentum after 2003, when it was first implemented by Walmart in their supply chain operations. When the technology was introduced, many industries were reluctant to adopt it, as it was neither cost effective nor robust enough to provide a good return on investment (ROI). Subsequent research and advancements in the technology have helped to reduce the price of installing RFID and wider applications. As a result, the impact is now being felt due to the realization of its efficiency, productivity and cost and time-saving, and the stage now is set for RFID applications in everyday life in many application areas.

RFID technology is being widely applied in retail and supply chain activities of fashion, textiles and many other consumer goods. IDTechEx forecasted that in 2022, the RFID market will be worth US$14.9 billion, and in 2030, US$30 billion. This cost includes the investment on RFID technology (including RFID tags, readers and software) for both passive and active RFID systems in all manufacturing and retail sectors. Hence, the future market of RFID technology is as bright as that of mobile and computer technology.

Looking at the huge potential, I was inspired to write this book on RFID technology. While I was doing the initial background research on RFID, I found some books focusing on the RFID technology and applications in supply chain. However, I was interested in the application of RFID technology in my own area of research on fashion and textiles, manufacturing and supply chain. A detailed search of the existing body of knowledge resulted in some journal publications in the area. However, there were no commercial publications available on the application of RFID technology in the field of fashion and textiles. Because of this prominent gap, I felt inspired to write a book on RFID technology. Hence, this book intensively focuses on the application of RFID technology on supply chain activities of fashion and textile, starting from fiber to finished garment and then to the consumers via retail.

The book "Radio Frequency Identification (RFID) Technology and Application in Fashion and Textile Supply Chain" is a result of my intensive work on RFID technology, which took about a year to complete. I collected the information from various peer-reviewed journals; trade magazines; websites of various companies, including fashion and textiles; information from various research organizations and RFID manufacturers; and the websites of fashion distributors. I also visited wool and cotton farms in Australia; and communicated with wool, silk and cotton growers in India and China to identify recent trends in RFID application in these areas. Surprisingly, many facts were revealed during the visit that are not known to many academics and researchers working in the area.

I have tried to make this book informative by covering all aspects of RFID applications in the fashion and textile supply chain. The early chapters describe the history of RFID technology and its systematic development. Later chapters describe the fundamental principles, components and workings of RFID systems. The role of existing barcode technology in fashion and textile technology are also discussed. Both barcode and RFID technologies are compared with respect to their fundamentals, advantages and disadvantages. The major applications of RFID technology in various processes are also summarized.

The application of RFID technology in production of various natural fibers such as wool, silk and cotton, and manufacturing of synthetic fibers such as polyester, nylon and viscose are discussed. My findings on the recent status of RFID technology in wool fiber growing, silk sericulture and cotton farming will be of special interest to readers. Similarly, the use of RFID technology in yarn manufacturing to detect bobbins, quality improvement and material management will also help readers to gain additional knowledge in the field. The application of RFID in managing the production, materials and storage during fabric manufacturing is also widely covered. Another important area of RFID use is chemical processing, which includes fabric dye batch identification, chemical and auxiliary management, labour management, and costing.

The applications of RFID in fashion retail and logistics operations provide valuable information to students, academics and business executives. This book covers more recent applications of RFID in retail such as smart shelves, smart fitting rooms, shrinkage reduction, automatic replenishment of inventory, point-of-sale (POS) supports and accuracy of inventory. The book has intensively studied the applications of RFID technology by leading global fashion brands and their future directions. The limitations of RFID technology and future directions are also discussed in the concluding chapter.

The book is written in simple language so that people with minimal knowledge in RFID technology can easily understand the subject matter. It is anticipated that this book, with its range of scientific knowledge, fundamental principles and application areas, will disseminate knowledge to students, academics, research communities and manufacturers working in fashion and textiles.

I extend my sincere thanks to all the experts who have helped to complete the chapters successfully. My special thanks to the authors and publishers for providing permission to use the images, tables and other information in the book. I am also thankful to the Taylor and Francis publication team for their support in making this book a reality. I hope this book will help readers in the field of fashion and textiles in acquiring relevant information on RFID Technology.

Dr. Rajkishore Nayak
Senior Lecturer, Fashion Merchandising
School of Communication and Design
RMIT University Vietnam

About the author

Dr. Rajkishore Nayak is currently working as a senior lecturer (Fashion Merchandising) at the School of Communication and Design, RMIT University, Vietnam. He completed his PhD in fashion and textiles from the School of Fashion and Textiles, RMIT University, Australia. He has around 18 years of experience in teaching and research related to Fashion and Textiles. His research areas include contemporary fashion manufacturing, fashion and textile sustainability; waste management and recycling; social media marketing in fashion; protective textiles and advanced materials.

He has published six books, 90 peer-reviewed journal papers and 20 conference proceedings. Dr. Nayak was awarded with the "RMIT University Research Excellence Team Award-2015." He also received the "RMIT University Teaching and Research Excellence Award-2012" and an "RMIT University International Scholarship-2008." He worked with the School of Fashion and Textiles, RMIT University, Australia from 2012–2016 in teaching and research before joining RMIT University Vietnam.

chapter one

Introduction to radio frequency identification

KEYWORDS: *RFID, fashion and textiles, supply chain, inventory management, retail management*

1.1 Introduction to RFID

The rapid pace at which technological innovations are being introduced in the world poses a potential challenge to retailers, suppliers and enterprises of many businesses. The approach to managing businesses has changed significantly with the application of the recent trend in new technologies. In the field of Information Technology (IT), there has been rapid growth in the last 30 years (Want, 2006; Landt, 2005). One of the most promising technological innovations in IT has been **R**adio **F**requency **Id**entification abbreviated as RFID (Nayak and Padhye, 2015; Bottani et al., 2009).

Radio Frequency Identification (RFID) technology was developed in 1945 as an espionage tool invented by Leon Theremin for the Soviet government (Nikitin, Parks and Smith, 2013; Tedjini et al., 2012). It has become an integral part of many organizations in supply chain management (Kwok and Wu, 2009), retail (De Marco et al., 2012), automatic toll collection (Khan, Yakzan and Ali, 2011), animal tracking (Floyd, 2015), ski lift access, library book tracking (Choi et al., 2006), theft prevention (Rekik, Sahin and Dallery, 2009), vehicle immobilizer systems (Jia et al., 2012), railway rolling stock identification (Ustundag and Tanyas, 2009), movement tracking (Brusey and McFarlane, 2009), security, healthcare (Yao, Chu and Li, 2010) and fashion and textile retail management (Nayak et al., 2007). RFID can make companies more competitive by changing the speed and accuracy of related processes in supply chain, logistics, manufacturing and retailing.

RFID is an electronic identification system, one of the most impactful innovations of the 20th century. RFID technology is based on wireless communication, using electromagnetic fields for automatic identification of an object, an animal or a person carrying RFID tags (Nath, Reynolds and Want, 2006). An RFID tag contains electronically stored information, which can be accessed by a reader when it is operated within the range of the tag. Researchers and developers spent several years creating a functional system that operates under similar principles to barcodes.

RFID technology has been rapidly adopted by many industries, including fashion and textiles.

Many of us use various RFID systems in our daily lives, without knowing they are based on RFID technology. Examples of RFID technology-enabled systems include: access cards to offices, train tickets, cards to operate lifts or even contactless car keys used to start a car. This technology can be extended to many production and distribution operations including fashion and textiles. The major driving factor for adaptation of RFID technology are efficiency in tracing and tracking; reduced time and labor cost; increased visibility; and accuracy of inventory level (Sari, 2010).

The RFID technology has been in use for the last 50 years or so, but it gained lots of attention after 2003, when it was first implemented by Walmart (Zelbst, Green and Sower, 2010). Hence, the stage was set for RFID application in everyday life in many application areas. In the initial period of its implementation, many industries assumed that the technology was neither cost effective nor robust enough to provide a good return on investment (ROI). However, the impact is now being felt due to the realization of its efficiency, productivity, cost and time saving.

The use of RFID technology is rapidly growing especially in retail and supply chain management. IDTechEx had forecasted that the total RFID market will be worth US$11.2 billion in 2017; it was US$10.52 billion in 2016 and US$9.95 billion in 2015 (Raghu, 2017). This cost included the investment on the technology (including RFID tags, readers and software) for both passive and active RFID systems in all manufacturing and retail sectors. In addition, IDTechEx forecasts that in 2022, the RFID market will be worth US$14.9 billion and in 2030 it will be worth US$30 billion. The growth of RFID technology is noted in Figure 1.1 based on sector and component.

The fashion and textile industries are looking for automation to increase efficiency and reduce lead time (Nayak and Padhye, 2017; Nayak and Padhye, 2018b). In the field of fashion and textile supply chain, there are many tangible benefits of implementing RFID. Many manufacturers and retailers are switching to RFID technology in the place of barcodes due to RFID's ability in automatic data capturing, tracking and automatic identification. RFID technology in combination with the electronic product code (EPC) can bring benefits that were never realized before (Nayak and Padhye, 2015).

Furthermore, we are on the verge of a fourth industrial revolution, also termed as "Industry 4.0". The first industrial revolution was coined in the 18th century, which started with the invention of the spinning Jenny, the mechanical loom; in addition to readily available steam technology. The second industrial revolution started in the early 20th century

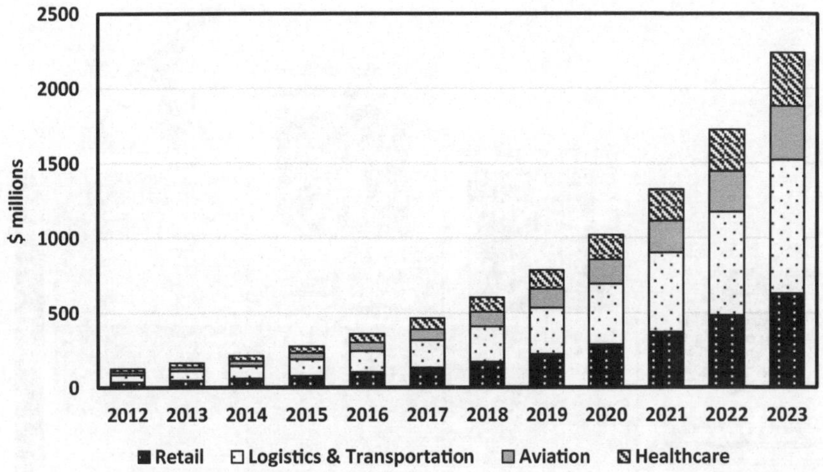

(a)

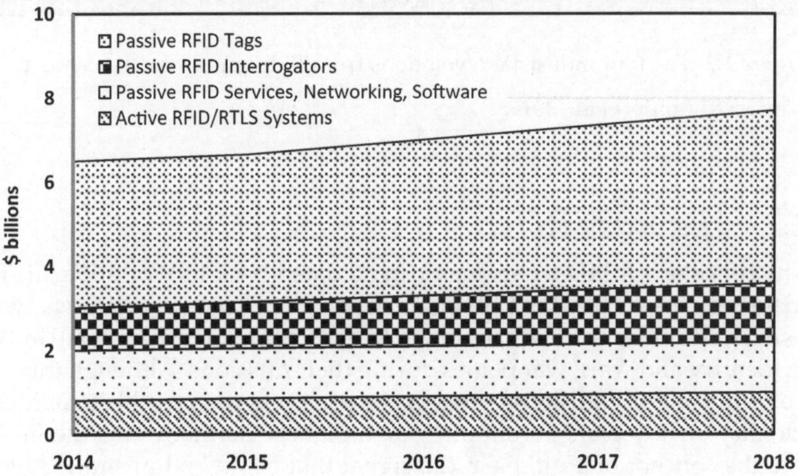

(b)

Figure 1.1 The growth of RFID technology: (a) based on sectors (in $ Millions) and (b) based on RFID systems and components.

Source: (a) ((Digital-Journal), no date) and (b) (Raghu, 2017).

with the wide availability of electricity, which focused on mass production driven by the Ford automotive company. The third industrial revolution started in the 1970s, which was focused on automation and computerization. The fourth industrial revolution has mainly focused on robotics, cyber-physical systems and IT integration. The industrial revolution is depicted in Figure 1.2 (Simonis et al., 2016).

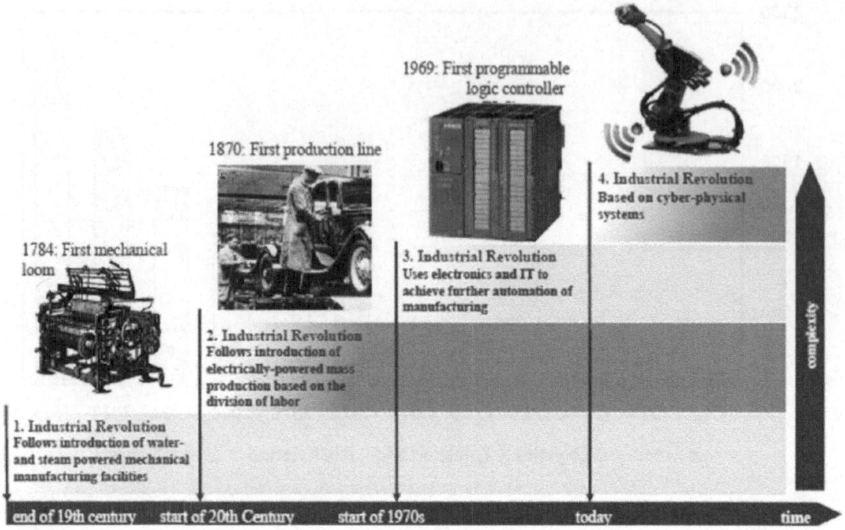

Figure 1.2 The four industrial revolutions from 18th century to 21st century.

Source: Simonis, et al., 2016.

1.2 History of RFID

The concept of RFID technology can be dated back to World War-II. The major powers such as the US, Britain, Japan and Germany in this war were using radio detection and ranging (RADAR) technology discovered in 1935 to identify and warn war planes during their missions. Although this was helpful to avoid collisions between planes, it was not possible to uniquely identify enemy planes from one's own planes. Germany was the first to use the concept of a unique radio signal that could be transmitted by its own planes as they returned to the base, and not by their enemy's planes. This was the first passive RFID system used in war planes, which was used to alert radar crew about their planes (Landt, 2005). The first active RFID system (known as IFF—i.e., identify friend or foe) was developed by Watson Watt in a secret project in Britain (Curran and Norrby, 2009). A transmitter was used in British planes, which would only respond to signals received from British radar stations.

These two principles—RADAR technology and radio signals invented during World War-II—are used in today's passive and active RFID systems. There was tremendous growth in radio communications in the second part of the 20th century (1950–2000), particularly for entertainment, which used radio waves. Due to the increased demand for radio, the term radio frequency (RF) was used and new jobs were created, such as RF Engineers.

Various technological inventions during this period, such as transistor, integrated circuit (IC), microprocessor and inventions in circuit etching, helped in the gradual development of RFID tags (Jones et al., 2005).

Furthermore, intensive research during 1950–1960s in the US, Europe and Japan helped to bring the RFID technology into commercial use in many fields. There have been reports from the Soviet Union that Leon Theremin designed a listening device in 1945, which used electromagnetic radio waves (Galeyev and Chudnovsky, 1996). The diaphragm of the device was used to alter the shape of the resonator using the vibration of the sound waves. The resonator in turn modulated the radio frequency. Although this device was used for listening, it resembled the passive RFID systems, which were activated by sound waves from external sources.

The fundamental concepts of RFID, how it works and its components are displayed in Figure 1.3, which is explained in Chapter 2 (Nayak and Padhye, 2015; Miles, Sarma and Williams, 2008). The major components of a RFID system are: RFID tag or transponder, antenna, reader or interrogator and the control system or computer. Although RFID components such as tags, antenna and reader were designed during the 1960s, the large size of the components was a major obstacle to commercial use. The launch of Microprocessor-4004 in 1971 by Intel Corporation was a milestone achievement in the design of RFID tags (Gelsinger, 2001). The microprocessor helped in the design of microchips and the concept was extended to the field of RFID in designing miniaturized RFID systems. The RFID tags designed subsequently were able to sense and store climatic conditions such as temperature and humidity with the inclusion of sensors and additional memory.

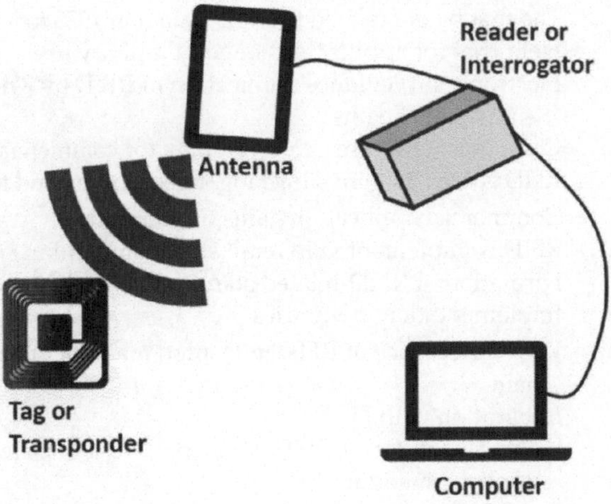

Figure 1.3 Components of a RFID system.

The initial research on RFID systems was based on passive tags, which subsequently switched to the active tags. The first US patent on an active RFID tag with rewritable memory was granted to Mario Cardullo on January 23, 1973. It was the basis of modern RFID tags (Shukla, Shah and Save, 2013). Prior to this, the passive RFID system consisted of a transponder with 16-bit memory. It was demonstrated in 1971 to the New York Port Authority (NYPA) for use in toll devices. Furthermore, a US patent was granted to Charles Walton in 1973 for a passive transponder used to unlock a door without a key (Walton, 1973). Subsequently, this technology was licensed to Schlage (a lock maker) and other companies. Table 1.1 shows the chronological developments in RFID technology.

In 1978, Tom Wilson and Matt Lezin successfully implanted RFID tags into cows on a dairy farm (Banks et al., 2007). The RFID system was used to monitor each cow's activities such as feeding habits, breeding cycle and other health issues. Since then, the use of RFID tags was diversified into other animals such as sheep, horses, goats and alpacas. The recent RFID tags can be implanted in the body of the animals that keep track of their behavior, feeding habits, movement, sleep patterns and other activities.

Table 1.1 Chronological development of RFID technology

Period	Development
1920–1940	Development of RADAR technology The concept of RFID coined prior to the World War-II
1941–1960	RFID technology used for identifying enemy airplanes The RFID technology that is in use today was invented in 1948 Laboratory work for commercial application of RFID begins
1961–1980	The first patent related to RFID issued in 1973 to Mario Cardullo Field trial for applications beyond military use Electronic surveillance application of RFID by Knogo and Sensormatic begins Companies and researchers prepare for commercial applications RFID is used in animal tracking, toll collection and tracking objects
1981–2000	Commercial application gains momentum RFID is implemented in retail stores and within supply chains Formation of RFID-related standards Implementation in libraries
2000 onwards	Wide acceptance of RFID in fashion manufacturing and supply chain Integration with IT Walmart, Tesco and US DoD (Department of Defense) declared suppliers' mandates Focus on miniaturization of the technology Efforts to reduce cost

The first application of RFID tags in shipping containers was adopted by the Department of Defense (DoD) in the United States (Xiao et al., 2006). It was adopted to achieve the benefit of tracking military logistics during war or peace and to fulfill the logistic shortfall. The RFID tags used in the containers were able to track both container movement as well as that of their contents. This helped the military to predict dates of delivery and make any necessary adjustments.

The concept of applying RFID tags in logistics is widely used by many global logistics service providers (LSPs) including fashion. The technologies in recent RFID systems are quite advanced, which is possible due to developments in computer and IT in addition to developments in radio communication. RFID has gained wide acceptance in areas such as order fulfillment, warehouse management, inventory control, production planning and tracking activities. Although barcode technology is still predominant in many retail sectors, the benefits of using RFID tags are being realized by many industries.

In the retail sector, Walmart announced in 2003 that its top 100 suppliers should use RFID tags in its containers by the end of 2006 (Weinstein, 2005). After this announcement, many other retail stores such as Best Buy followed the footprint of Walmart and began using RFID tags. Other enterprises soon followed the trend and implemented RFID technology.

Both the RFID technology and barcode systems are based on the concept of identifying objects (White et al., 2007; Dos Santos and Smith, 2008). Each object in the supply chain carries its own unique identification code, which can be accessed by suppliers, manufacturers and retailers. The instant access to the information can help to monitor work progress, estimate the delivery date and plan the date for launching a new product. However, RFID enhances data processing and works as a complementary technology to barcode.

RFID is a more advanced system than barcode technology in terms of non-optical proximity communication, information density and two-way communication ability (Nath, Reynolds and Want, 2006; Roberts, 2006; Miles, Sarma and Williams, 2008). Recent developments in microelectronics and data processing have enabled the use of less expensive and smaller components with improved efficiency in RFID technology. Hence, the application of RFID in various areas particularly in supply chain has witnessed tremendous growth (Bagchi et al., 2007). Recent research on RFID has focused on inventory management, logistics, transportation, manufacturing, asset tracking and object location (Garcie et al., 2007; Sun, 2012; Nayak and Padhye, 2014; Piramuthu, Wochner and Grunow, 2014;).

The World Trade Organization (WTO) is continuously trying to facilitate international trade by reducing barriers among countries.

This in turn has resulted in the shifting of fashion and textile manufacturing to the countries where labor cost is cheap. Christopher, Lowson and Peck (2004) characterize the fashion industry by wider collection/more choices, seasonally impacted, variable and short life-cycles, complex supply chain and distribution activities. Profitability in fashion industries is going down due to the competitive nature of business, thus requiring better coordination between all partners in the supply chain. Fashion companies that have previously dominated the market at some stage have lost market share to emerging companies like Zara, H&M and Benetton, as the business strategy of those companies has involved more inventory turns. They also offer a wider range of fashion styles in short span giving rise to the concept of "Fast Fashion". All these changes in the fashion industry need to use new technologies widely in order to reduce lead time and cost and most importantly, to efficiently manage inventory. Among new technologies, RFID is of particular interest as it has the capability of providing solutions to some of these issues (Moon and Ngai, 2008).

1.3 Why RFID

RFID technology has been widely adopted by many manufacturing, retail and logistics industries. Many organizations are following in the footsteps of Walmart and the Department of Defense (DoD), which were the first to include the technology in their supply chain. Walmart mandated in 2003 that its top suppliers use RFID technology for pallets and item level tracking with the help of EPC codes (Weinstein, 2005). Similarly, the DoD mandated to its top 100 suppliers to use its RFID system for item level tracking (Wu et al., 2006).

Walmart and DoD are together the world's largest retail operator and the world's largest supply chain operator, respectively (Van Wassenhove, 2006; Roberts and Berg, 2012). With the increase in supply chain activities and the decreasing cost of implementation, RFID technology has increasingly been adopted into their mainstream operations. With the emergence of international standards such as the EPC class 1 G2 (Generation 2) protocol and ISO 18000-6, RFID technology is moving towards standardization (Wu et al., 2013). Walmart and DoD have found the technology quite helpful to achieve their business objectives.

The major driving factors for the use of RFID technology are the accuracy of information, availability of real-time information, automatic information update, managing stock levels, reduced time, non-contact technology, increased speed, reduced labor cost, product visibility and reduced shrinkage (Whitaker, Mithas and Krishnan, 2007; Tan, 2008). Due to these benefits, RFID technology is being widely used in major areas of application, which includes asset management and supply chain activities. This technology has been widely accepted by a wide range of industries including fashion retail.

Fashion products are characterized by short life cycles, volatility in demand, large product varieties, customized requirements and high-quality requirements (Nayak and Padhye, 2015). Apparel industries always focus on the sourcing of low-cost material and cheap labor. Furthermore, finished fashion products should also be available at cheaper prices, which makes it hard for apparel manufacturing industries to survive in developed countries (Nayak and Padhye, 2014). To become cost competitive, many apparel industries are seeking to optimize the cost structure in their supply chains by adopting artificial intelligence (AI), automation and new technologies such as RFID (Nayak and Padhye, 2017; Nayak and Padhye, 2018a). Several fashion brands have followed in the footsteps of Walmart and DoD in implementing RFID technology.

The other reason for wide acceptance of RFID can be attributed to globalization of fashion manufacturing. The stiff competition from developing countries, increasing labor prices and stricter regulations in developed countries have resulted in the relocation of fashion manufacturing from developed countries to developing countries (Nayak and Padhye, 2015). In developing countries, the fashion manufacturers face cutthroat competition and want to be ahead in the race by reducing the cost of manufacturing mainly by focusing on the adoption of new technologies such as RFID and the elimination of manual work through automation.

Although many apparel manufacturing and retail stores have undergone modernization with new technologies and automation, they still suffer from the following challenges:

- Manual and inaccurate ways of inventory management
- Manual, time-consuming and expensive methods of checking physical inventory
- High amount of shrinkage in the supply chain process, including retail
- Inaccurate and delayed information availability, making supply chain activities difficult to manage
- Out-of-stock inventories that lead to lost sales and reduced number of customers
- Difficulty in tracking and tracing of products in the supply chain
- Missing shipments that lead to customer dissatisfaction
- Low productivity and efficiency due to manual methods of doing work

RFID technology is being increasingly adopted by many fashion brands due to the reasons discussed above and will be the future technology for many fashion brands. The use of RFID technology can help to a large extent to eliminate the drawbacks of manual processes as mentioned above.

1.4 Application areas

RFID technology can be applied in many manufacturing industries such as civil, electronic, automotive, food, logistics, healthcare, tourism, library, textile, aviation and fashion (Kärkkäinen, 2003; Jones et al., 2005). RFID technology has been used in a range of applications with the most important applications as listed in Table 1.2. The application areas are also shown in Figure 1.4.

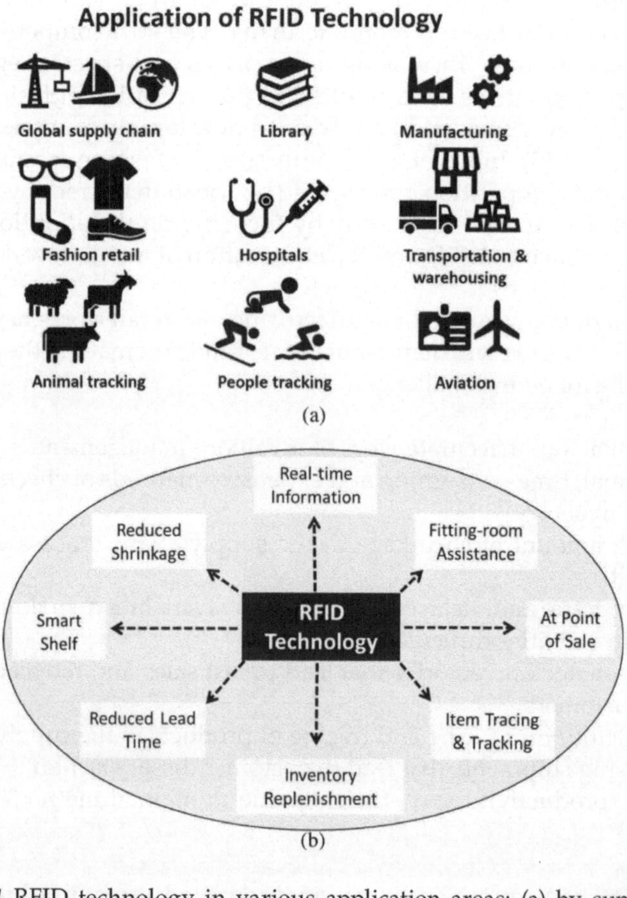

Figure 1.4 RFID technology in various application areas: (a) by supply chain activity; and (b) by point of application.

Table 1.2 Various applications of RFID technology

Application areas	Description
Retail stores	There is great potential for RFID technology in retail stores. The technology can be used for product tracking, merchandise replenishment, theft prevention, customer relationship management (CRM), managing returns and counterfeit prevention. At the point-of-sale, RFID can be used for self-checkout by customers. It can also be used for the management of customer returns or reverse logistics.
Assets tracking	Tracking various assets in large organizations such as hospitals and corporate offices is a challenging task. RFID technology can be used to track immovable assets such as wheelchairs, beds and storage racks as well as in-motion assets such as laptops, cellphones or vehicles. Similarly, tracking the exact location of assets that are in continuous motion in warehouses (such as forklifts and pallet jacks) is a daunting task. The use of RFID technology can trace the exact location of tools and equipment, which increases ROI. One recent application includes tracking high-valued engine components during aircraft manufacturing or tracking aircraft during testing or test-runs. It can also be used to track parked vehicles, rail coaches and shipping containers.
Animal tracking	The movement of different animals can be tracked using RFID tags. This can help in studying the life pattern and movement of different animals. Many zoos and animal protection groups are using this technology to study the life pattern of various animal species.
People tracking	RFID is used in jails and hospitals to identify criminals and patients, respectively. Tracking jail inmates is widely used in many jails in the US to closely monitor the movement of inmates. In hospitals, RFID technology can be used to identify and monitor patients requiring special care, emergency patients and people suffering from anxiety.
Tracking newborns & infants	The swapping of newborns is a common problem faced in hospitals in many countries, which is an emotional stress for many parents. Therefore, many hospitals have looked into various solutions to resolve these issues and found that RFID technology is a solution for this problem. Both the baby and the mother are fixed with secure wristbands fitted with RFID tags. A swapping of babies will be automatically detected by the RFID system and will alert the hospital staff and parents.

(Continued)

Table 1.2 Various applications of RFID technology *(Continued)*

Application areas	Description
Product tracking	The RFID system can be used in the manufacturing industry to track the progress of different products. Any process delay or bottlenecking can be identified if the product is not progressing. During supply chain of the products, they can be easily identified as per their location, which can be used to estimate the arrival date.
Supply chain management	RFID is used for the tracking of various individual products as well as tracking the shipping containers. It is used for inventory management and automatic updating of stock. It can also be used for shrinkage prevention in the supply chain process.
Libraries	RFID technology can be used in libraries for book management and prevention of theft. It can also be used to check the availability of books or the correct location of the books within the library, if they are moved.
Toll collection	RFID technology is used for toll collection on toll roads, which is the most common use by millions of people. The vehicles using RFID technology do not need to queue up for toll payment. The vehicles use a small RFID-enabled device in the windshield. When a vehicle with an RFID-enabled device approaches in the fast lane, overhead readers in the toll collection center read the details of the vehicle and process the payment.
Preventing counterfeit	There has been tremendous growth in counterfeiting luxury fashion products. RFID technology can prevent counterfeiting. The data or logo relating to the brand can be stored in an RFID chip, which can be read by an RFID reader by its radio frequency. An internal factory-programmed ID that can't be duplicated is used to store the information to avoid the counterfeit.
Office management	RFID tags can be used to manage the employees of an office. Each employee can be provided with a unique card with RFID technology. Hence, the working hours and the number of days they work can be managed by the RFID tags.
Immigration management	RFID tags can also be used to manage the immigration of each person, including passport and visa management.
Sensing and monitoring	RFID tags can be attached with other sensors used for sensing and monitoring temperature, pressure, humidity and volume. Hence, the products needing specific conditions during transportation can be closely monitored if the conditions are appropriate. Similarly, suitable sensors can be attached with RFID tags to monitor people's heart rate and blood pressure.

1.5 Scope of the book

To date, there are some book publications on RFID applications in various manufacturing, retail and supply chain operations. However, it was found by the author that no publication covers a detailed application of RFID technology in fashion supply chain from fiber production to fashion retailing. Therefore, the author focused on updating the existing body of knowledge in an innovative manner by doing a thorough background research from the fibre stage till the products reach to the consumers, which can help the readers to gain new knowledge in the field. Furthermore, the author has included new knowledge from his own working experience in the field for about two decades.

This book can be used as a textbook in universities where RFID is studied; it can provide guidelines to those in the fashion industry implementing these new technologies; it can help supply chain operators select the right type of RFID technology; and it can help technology developers understand the demands and requirements of the current fashion market. The author has given a brief description of each chapter in the following section.

Chapter 1 provides a brief introduction to RFID technology. This chapter provides a brief history of the RFID technology and describes the chronological developments in the field. The major reasons for wide acceptance of RFID technology are also discussed, as is the application of RFID in various areas. Furthermore, this chapter gives a highlight of the contents of the book.

Chapter 2 compares RFID technology with the existing barcode technology, which is widely used in many businesses including fashion and textiles. A brief history of barcode technology is discussed. It compares both identification systems in terms of technology, working principles, cost, benefits, advantages and disadvantages. Furthermore, the working principles of RFID technology and the components and types of RFID systems are also discussed.

Chapter 3 discusses the application of RFID technology in textile manufacturing processes, including fiber production and yarn manufacturing. Applications of RFID in plant fibers, animal fibers and synthetic fiber production, including bale transportation and fiber testing management, are discussed in this chapter. This chapter also discusses RFID applications in yarn manufacturing such as bobbin management, yarn quality tracking, materials management and machine maintenance.

Chapter 4 discusses RFID applications in fabric manufacturing and fabric chemical processing. Various applications in fabric manufacturing such as material management, machine maintenance, production and productivity management, fabric quality inspection and labor management are discussed in this chapter. Furthermore, various applications of

RFID technology in chemical processing are discussed in this chapter, which is one of the most important value-added steps in textile industries. Applications include managing chemicals, routing process path, batch identification, increased visibility and labor management.

Chapter 5 discusses the application of RFID in fashion manufacturing. Various applications in garment manufacturing, starting from design development to final inspection, are covered in this chapter. This chapter also discusses in detail RFID technology in fabric spreading, inventory management, resource allocation, production control, machine maintenance, labor management, quality management, warehouse management and tracking of products.

Chapter 6 discusses the application of RFID technology in fashion logistics activities. Logistics activities play an important role in the success of many fashion businesses. Applications of RFID in fashion logistics include: warehouse management, inventory management, managing loading and unloading, identification of products, replenishment of stock and reducing shrinkage. The benefits of using RFID technology in fashion and textile supply chain are also discussed.

Chapter 7 discusses the application of RFID technology in fashion retail, where the consumers purchase the fashion. Various applications of RFID, including product tracking, retail management, merchandise replenishment, information management, shrinkage reduction, theft prevention, brand segregation, consumer self-checkout and managing returns are included in this chapter. In addition, customer relationship management (CRM) and RFID-based care labelling systems are discussed in this chapter.

Chapter 8 discusses case studies of fashion brands in the application of RFID. Different fashion brands that have already adopted RFID technology in their manufacturing and supply chain processes are discussed, including the benefits achieved. Similarly, fashion brands that are currently planning to adopt the technology are also discussed.

Chapter 9 discusses the challenges and future directions of RFID technology. Various challenges faced by RFID technology such as high cost, security and privacy issues; RFID attacks by hackers; cloning of tags; compatibility issues; information tracking; and data management are all discussed in this chapter. Furthermore, ethical issues, lack of standardization, research on RFID and future directions ae also discussed in this chapter.

1.6 Conclusions

RFID technology is one of the most revolutionary technological developments of information technology in the 21st century. The application of RFID technology is rapidly growing in manufacturing industries and supply chain activities since it was mandated by Walmart and DoD in 2003. RFID has been widely used in various applications such as people tracking; tracking of babies and products; supply chain management; library management; toll collection; counterfeit prevention; office management; immigration management; and sensing and monitoring. This chapter has discussed the history of RFID technology and the reasons for its continual growth. This chapter has also discussed potential applications of the technology. At present, RFID technology is suffering from many drawbacks such as high cost and lack of standardization, which are the major impediments in its adoption. In future, it will be widely accepted by many sectors once the drawbacks have been resolved.

References

Baggchi, U., Guiffrida, A., O'Neill, L., Zeng, A. & Hayya, J. 2007. The effect of RFID on inventory management and control. In: Jung H., Jeong B., & Chen F. F. (Eds) *Trends in Supply Chain Design and Management*. Springer Series in Advanced Manufacturing. Springer, London. Springer.

Banks, J., Pachano, M. A., Hanny, D. & Thompson, L. G. 2007. *RFID Applied*, John Wiley & Sons.

Barcode-Graphics. 2018. *What is RFID?* [Online]. Available: https://www.epc-rfid.info/rfid [Accessed 15 August 2018].

Bottani, E., Ferretti, G., Montanari, R. & Rizzi, A. 2009. The impact of RFID technology on logistics processes of the fashion industry supply chain. *International Journal of RF Technologies: Research and Applications*, 1, 225-252.

Brusey, J. & McFarlane, D. C. 2009. Effective RFID-based object tracking for manufacturing. *International Journal of Computer Integrated Manufacturing*, 22, 638-647.

Choi, J.-W., Oh, D.-I. & Song, I.-Y. 2006. R-LIM: an affordable library search system based on RFID. International Conference on Hybrid Information Technology (ICHIT). IEEE, 103-108.

Christopher, M., Lowson, R. & Peck, H. 2004. Creating agile supply chains in the fashion industry. *International Journal of Retail & Distribution Management*, 32, 367-376.

Curran, K. & Norrby, S. 2009. RFID-enabled location determination within indoor environments. *International Journal of Ambient Computing and Intelligence*, 1, 63-86.

De Marco, A., Cagliano, A. C., Nervo, M. L. & Raffle, C. 2012. Using System Dynamics to assess the impact of RFID technology on retail operations. *International Journal of Production Economics*, 135, 333-344.

Digital-Journal. 2016. *Printed and Chipless RFID Market: Global Industry Analysis Report, Share, Size, Growth, Price Trends and Forecast 2016-2023* [Online]. Available: http://www.abnewswire.com/pressreleases/printed-and-chipless-rfid-market-global-industry-analysis-report-share-size-growth-price-trends-and-forecast-2023-fractoviaorg_91114.html [Accessed 18 August 2018].

Dos Santos, B. L. & Smith, L. S. 2008. RFID in the supply chain: panacea or Pandora's box? *Communications of the ACM*, 51, 127-131.

Floyd, R. E. 2015. RFID in animal-tracking applications. *IEEE Potentials*, 34, 32-33.

Galeyev, B. M. & Chudnovsky, V. 1996. Light and shadows of a great life: in commemoration of the one-hundredth anniversary of the birth of Leon Theremin, pioneer of electronic art. *Leonardo Music Journal*, 6, 45-48.

García, A., Chang, Y., Abarca, A. & Oh, C. 2007. RFID enhanced MAS for warehouse management. *International Journal of Logistics*, 10, 97-107.

Gelsinger, P. P. 2001. Microprocessors for the new millennium: Challenges, opportunities, and new frontiers. International Solid-State Circuits Conference, 2001. Digest of Technical Papers. IEEE, 22-25.

Jia, X., Feng, Q., Fan, T. & Lei, Q. 2012. RFID technology and its applications in Internet of Things (IoT). International Conference on Consumer Electronics, Communications and Networks (CECNet). IEEE, 1282-1285.

Jones, P., Clarke-Hill, C., Hillier, D. & Comfort, D. 2005. The benefits, challenges and impacts of radio frequency identification technology (RFID) for retailers in the UK. *Marketing Intelligence & Planning*, 23, 395-402.

Kärkkäinen, M. 2003. Increasing efficiency in the supply chain for short shelf life goods using RFID tagging. *International Journal of Retail & Distribution Management*, 31, 529-536.

Khan, A. A., Yakzan, A. I. E. & Ali, M. 2011. Radio frequency identification (RFID) based toll collection system. Third International Conference on Computational Intelligence, Communication Systems and Networks (CICSyN). IEEE, 103-107.

Kwok, S. K. & Wu, K. K. 2009. RFID-based intra-supply chain in textile industry. *Industrial Management & Data Systems*, 109, 1166-1178.

Landt, J. 2005. The history of RFID. *IEEE Potentials*, 24, 8-11.

Miles, S. B., Sarma, S. E. & Williams, J. R. 2008. *RFID Technology and Applications*, Cambridge University Press, Cambridge.

Moon, K. & Ngai, E. 2008. The adoption of RFID in fashion retailing: a business value-added framework. *Industrial Management & Data Systems*, 108, 596-612.

Nath, B., Reynolds, F. & Want, R. 2006. RFID technology and applications. *IEEE Pervasive Computing*, 5, 22-24.

Nayak, R., Chatterjee, K., Khurana, G. & Khandual, A. 2007. RFID: Tagging the new era. *Man-Made Textiles in India*, 50, 174-177.

Nayak, R. & Padhye, R. 2014. Introduction: the apparel industry. *In*: Nayak, R. & Padhye, R. (eds.) *Garment Manufacturing Technology*. Elsevier.

Nayak, R. & Padhye, R. 2015. *Garment Manufacturing Technology*, Cambridge, UK, Woodhead Publishing.

Nayak, R. & Padhye, R. 2017. *Automation in Garment Manufacturing*, Cambridge, UK, Woodhead Publishing.

Nayak, R. & Padhye, R. 2018a. Artificial intelligence and its application in the apparel industry. *In*: Nayak, R. & Padhye, R. (eds.) *Automation in Garment Manufacturing*. Elsevier.

Nayak, R. & Padhye, R. 2018b. Introduction to automation in garment manufacturing. *In*: Nayak, R. & Padhye, R. (eds.) *Automation in Garment Manufacturing*. Elsevier.

Nayak, R., Singh, A., Padhye, R. & Wang, L. 2015. RFID in textile and clothing manufacturing: technology and challenges. *Fashion and Textiles*, 2, 1-16.

Nikitin, P. V., Parks, A. & Smith, J. R. 2013. RFID-Vox: a tribute to Leon Theremin. *Wirelessly Powered Sensor Networks and Computational RFID*. Springer.

Piramuthu, S., Wochner, S. & Grunow, M. 2014. Should retail stores also RFID-tag 'cheap' items? *European Journal of Operational Research,* 233, 281-291.

Raghu, D. 2017. *RFID forecasts, players and opportunities 2017-2027* [Online]. Available: https://www.idtechex.com/research/reports/rfid-forecasts-players-and-opportunities-2017-2027-000546.asp [Accessed 10 July 2018].

Rekik, Y., Sahin, E. & Dallery, Y. 2009. Inventory inaccuracy in retail stores due to theft: An analysis of the benefits of RFID. *International Journal of Production Economics,* 118, 189-198.

RFID-World. 2018. *The power in using RFID in the apparel industry* [Online]. Available: https://www.rfidworld.ca/author/jimhill [Accessed 15 August 2018].

Roberts, B. & Berg, N. 2012. *Walmart: Key Insights and Practical Lessons from the World's Largest Retailer,* Kogan Page Publishers.

Roberts, C. M. 2006. Radio frequency identification (RFID). *Computers & Security,* 25, 18-26.

Sari, K. 2010. Exploring the impacts of radio frequency identification (RFID) technology on supply chain performance. *European Journal of Operational Research,* 207, 174-183.

Shukla, S., Shah, S. & Save, P. 2013. RFID Based attendance management system. *International Journal of Electrical and Computer Engineering (IJECE),* 3, 784-790.

Simonis, K., Gloy, Y. & Gries, T. INDUSTRIE 4.0-Automation in weft knitting technology. IOP Conference Series: Materials Science and Engineering, 2016. IOP Publishing, 012014.

Slideplayer. 2016. *The evolution of RFID Technology* [Online]. Available: https://slideplayer.com/slide/10638805/ [Accessed 10 August 2018].

Sun, C. 2012. Application of RFID technology for logistics on internet of things. *AASRI Procedia,* 1, 106-111.

Tan, H. The application of RFID technology in the warehouse management information system. International Symposium on Electronic Commerce and Security, 2008. IEEE, 1063-1067.

Tedjini, S., Perret, E., Vena, A. & Kaddour, D. 2012. Mastering the electromagnetic signature of chipless RFID tags. In Karmaka, N.C. (eds.). *Chipless and Conventional Radio Frequency Identification: Systems for Ubiquitous Tagging. IGI Global,* 146-174.

Ustundag, A. & Tanyas, M. 2009. The impacts of radio frequency identification (RFID) technology on supply chain costs. *Transportation Research Part E: Logistics and Transportation Review,* 45, 29-38.

Van Wassenhove, L. N. 2006. Humanitarian aid logistics: supply chain management in high gear. *Journal of the Operational research Society,* 57, 475-489.

Walton, C. 1973. Electronic identification & recognition system. US patents: US3816709A.

Want, R. 2006. An introduction to RFID technology. *IEEE Pervasive Computing,* 5, 25-33.

Weinstein, R. 2005. RFID: A technical overview and its application to the enterprise. *IT professional,* 7, 27-33.

Whitaker, J., Mithas, S. & Krishnan, M. S. 2007. A field study of RFID deployment and return expectations. *Production and Operations Management,* 16, 599-612.

White, G., Gardiner, G., Prabhakar, G. P. & Abd Rajak, A. 2007. A comparison of barcoding and RFID technologies in practice. *Journal of Information, Information Technology and Organizations,* 2, 119-132.

Wu, H., Zeng, Y., Feng, J. & Gu, Y. 2013. Binary tree slotted ALOHA for passive RFID tag anticollision. *IEEE Transactions on Parallel and Distributed Systems,* 24, 19-31.

Wu, N.-C., Nystrom, M., Lin, T.-R. & Yu, H.-C. Challenges to global RFID adoption. Technology Management for the Global Future, 2006. PICMET 2006, 2006. IEEE, 618-623.

Xiao, Y., Shen, X., Sun, B. & Cai, L. 2006. Security and privacy in RFID and applications in telemedicine. *IEEE Communications Magazine,* 44, 64-72.

Yao, W., Chu, C.-H. & Li, Z. The use of RFID in healthcare: Benefits and barriers. IEEE International Conference on RFID-Technology and Applications (RFID-TA), 2010. IEEE, 128-134.

Zelbst, P. J., Green JR, K. W. & Sower, V. E. 2010. Impact of RFID technology utilization on operational performance. *Management Research Review,* 33, 994-1004.

chapter two

Overview of barcode and RFID technology

KEYWORDS: *barcode, RFID tag, interrogator, cost, security*

2.1 Introduction

A barcode is a representation of data or product information using parallel lines of varying width (Sriram et al., 1996; Hong-Ying, 2009). In other words, barcodes are data encoded in image form consisting of parallel lines. The image is comprised of high contrast regions such as solid black lines with white background. The barcodes work on the principle of symbology, which includes the extraction of data using symbolic images. Barcode technology is simple and inexpensive and therefore widely used in many operations such as airline luggage, secure mail, parcels, retail stores, tickets, cinemas, theaters and transportation (Akeroyd, 2010; Sriram et al., 1996).

The barcodes used in fashion retailing contain information on the product type, style, color and price (Basker, 2012). The barcodes can be one- or two-dimensional that can be read by an optical reader or scanner. The initial barcodes were one-dimensional, linear bars arranged in parallel manner. The development of computer technology has led to the use of two-dimensional barcodes, which uses various geometric patterns such as dots, rectangles and hexagons.

On the other hand, the Radio Frequency Identification (RFID) system is a contactless technology based on the principles of electromagnetic waves (Nayak et al., 2015). The information is stored in RFID tags, which are memory chips of varying capacity. A reader is needed to extract information from RFID chips, in addition to the hardware and software tools needed to integrate the RFID function (Nayak et al., 2007a). The reader can read the information from a distance depending on the range of operations of the RFID system.

Barcode technology has been operational for several years and it is well established. However, the recent trend is to switch from barcode technology to RFID technology due to the additional benefits of RFID technology (Nayak and Padhye, 2017; Nayak and Padhye, 2018). This chapter discusses the basics, history and working of barcode technology. Similarly, the working of RFID technology is also discussed in this chapter. Furthermore, barcode and RFID technologies are compared.

2.2 Barcode technology

Barcode technology has become a ubiquitous part of many modern businesses. Most products in retail stores are sold with a unique barcode. Similarly, barcodes are widely used in the healthcare sector, hospitals, manufacturing industries and supply chain management. The use of barcodes can help to keep the inventory record and reduce theft from retail outlets. Barcodes also help to manage customers with unique barcode card and sales management tools. This section will review the history of barcode technology and its working.

2.2.1 History of barcode

The initial work on barcode technology was started in 1948 by Bernard Silver and his friend Norman Joseph Woodland at the Drexel Institute of Technology in Philadelphia, USA (Kippelen and Durgin, 2010; Wikipedia, 2018c). This work was relating to a food chain's request to the Dean for a system to automatically read production information during check out. Both Bernard and Norman worked on different systems and came up with a system based on ultraviolet (UV) ink. However, it was not successful, as it was expensive, and the ink faded easily after few days.

Norman worked in his father's house to further develop the system after leaving Drexel Institute of Technology and was successful. Getting inspiration from the Morse code, he formed the first barcode using the sand on a beach. The concept was to use lines of varying widths extended linearly, which were read by a lightbulb and a photomultiplier tube (as with a movie projector). The subsequent design was based on narrow and wide circles instead of lines, which could be read from any direction. Both Norman and Bernard were granted a patent in 1952 (US Patent 2,612,994), which described barcode systems (linear and circular) and readers (electronic and mechanical) (Nadeem and Velasco, 1993).

Further works by David Collins, the involvement of the Association of American Railroads (AAR), IBM, General Motors and the development of the Universal Product Code helped in the development and growth of the barcode system (Wikipedia, 2018c). The Sims Supermarket in Australia adopted the technology in 1979 for the first time. Later, it was adopted by the US Department of Defense (DoD) in 1981 for identifying different military products. Subsequently it spread into the manufacturing, distribution and retail sectors.

A barcode contains information relating to the product, product type, manufacturer and type of industry, as discussed in Table 2.1.

Table 2.1 Description of different sections of a barcode

Codes (Bar, number or character)	What it indicates?
Left Hand Guard Bars	These bars work as a starting reference point for the scanner (Refer Figure 2.2).
Number System Character	This digit identifies manufacturer type or how the barcode will be used. For example, digit 3 is used for health and drug manufacturing industries, whereas 0, 6 and 7 are used in retail.
Number System Bars	These bars correspond to the Number System Character.
Manufacturer ID Number	Each company must apply for a Universal Product Identification Number with the Uniform Code Council (UCC). The UCC assigns each company a unique six-digit identification number for use on all of their products. The number is composed of the Number System Character and a five-digit manufacturer's code.
Manufacturer ID Bars	These bars correspond to the Manufacturer's ID Number.
Tall Center Bar	These bars serve as a middle reference point for scanning devices.
Item Number	Each company is responsible for assigning a unique five-digit number to each of their products.
Item Bars	These bars correspond to the Item Number.
Modulo Check Character	This digit is derived from a mathematical formula based on the unique set of numbers in each barcode that helps ensure the accuracy of the data scan.
Modulo Check Bars	These bars correspond to the Modulo Check Character.
Right Hand Guard Bars	These bars serve as an ending reference point for scanning devices.

Source: (Barcodeart, n.d.).

2.2.2 *Working of barcode*

A barcode system mainly comprises of three parts: (a) the printed barcode, (b) scanner and (c) computer system to read and extract the information.

2.2.2.1 *Printed barcode*

The first barcode created by Norman was in a bulls eye shape (Longacre, 2004). Later, it was modified to different shapes primarily printed on paper to suit the purpose. A printed barcode is the representation of data in machine readable and optical format. In a standard format, barcodes are printed with the use of parallel lines of varying width and contrasting sections between two lines. The lines and contrasting sections of a barcode can be created by several ways. More recent barcodes come in two different formats: (a) one-dimensional (1D) and (b) two-dimensional (2D).

The 1D barcodes come in different symbology formats such as code 11, code 39, code 49, code 93, code 128, EAN 2, EAN 5, UPC A and UPC E. Similarly, 2D barcodes come in different formats such as Aztec code, data matrix, dot code, Maxicode, PDF 417, shot code, SPARQ code, AR code and quick response (QR) format (Katona and Nyúl, 2013; Wikipedia, 2018a). Some of the 1D and 2D barcodes commonly used recently are shown in Figure 2.1.

Both the 1D and 2D barcodes contain some of the following information from a set of data such as: manufacturer ID, item number, country of origin, type of product and year of manufacturing. The information details for 1D and 2D barcodes are explained in Figure 2.2.

2D barcodes, which consist of square or rectangular cells in black and white, are rapidly replacing 1D barcodes (see Figure 2.2). 2D barcodes can store large amounts of data such as lot and serial numbers, product information, etc., which is not possible by the 1D barcodes. Furthermore, 2D barcodes can be scanned easily on curved surfaces, which is difficult for 1D barcodes. 2D barcodes are better for preventing counterfeiting and are more efficient in providing information and tracking the products. Hence, 2D barcodes are replacing 1D barcodes in many items from water bottles to airplane tickets and they are the future of barcode systems.

Figure 2.1 Different types of barcodes: (a) code 11, (b) UPC code, (c) Aztec code and (d) QR (Quick Response) code.

Source: (Wikipedia, 2018a).

(a)

(b)

Figure 2.2 Descriptions of information contained in a barcode: (a) 1D barcode and (b) 2D barcode.

Source: (a) (Barcodeart, n.d.) and (b) (Wikipedia, 2018b).

Due to their structure, 2D barcodes can retain more information than 1D barcodes of the same dimensions. The data are encoded both horizontally and vertically consisting of several patterns. 2D barcodes are free from alpha-numeric characters, but consist of dots, squares, circles, hexagons and other geometric patterns. These patterns can be related to the product information, website address, audio files or other binary data. 2D barcode scanners used to read the barcodes are different from 1D barcode scanners. Some of the recent smartphones can also read them. Some 2D barcode scanners are also compatible with 1D barcodes.

2.2.2.2 Scanner

Scanners are used to scan the barcodes to collect the information stored in the barcodes (Yaqoob and Riza, 2004). Scanners are handheld devices that work on different types of lighting sources, which are discussed below:

1. Optical scanner: Optical scanners use a camera to capture the image in the barcode (Figure 2.3(a)). The cameras can work in any orientation to capture the image, which is then transferred to the computer system for processing and extraction of information.

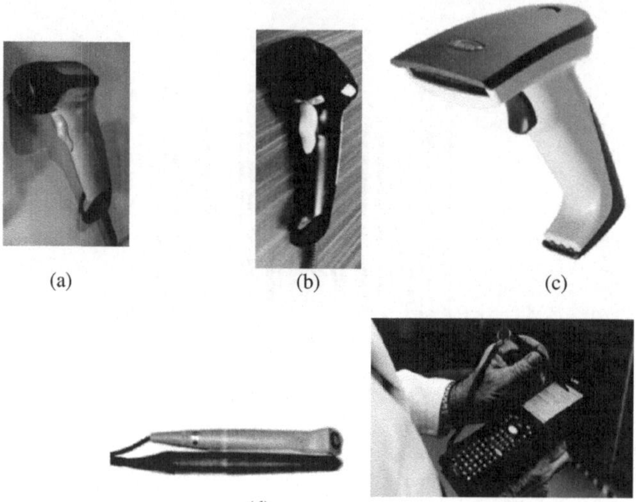

(a) (b) (c) (d) (e)

Figure 2.3 Various types of scanners: (a) Optical scanner, (b) Laser scanner, (c) CCD scanner, (d) Pen wand and (e) Slot scanner.

Source: (a) and (b): photo by Nguyen Tran Minh Chau; (c), (d) and (e) Attributed from en:wikipedia, photo by Argox.

2. Laser scanner: Many barcode scanners work on a laser scanner. Laser scanners use a vibrating mirror to sweep the laser's focal point across the whole width of the barcode as shown in Figure 2.3(b). The reflected light is received by the laser scanner. As the darker regions of the barcode receive higher amount of laser, the reflection will be less. This reflection is converted to digital image and compared with the stored information.
3. Charge coupled device or CCD scanner: The principle behind a CCD scanner is similar to an optical scanner. CCD scanners have several LED (light emitting diode) lights arranged in a long row that capture the image. They scan the image at a faster rate than optical scanner, but the range of the scan is limited (Figure 2.3(c)). The CCD scanners count the number of photons the barcodes emit back (i.e., the light vs the dark part of the barcode).
4. Pen wand: Pen wand is the simplest type of scanner; it is like a pen (Figure 2.3(d)). The pen wand with a lighting device is scanned from one end of the barcode to the other. The image is scanned by the pen wand and sent to the computer. However, it is essential to move the pen at a constant rate for accuracy of information.
5. Slot scanner: The slot scanner (Figure 2.3 (e)) operates under a similar principle as the pen wand. The barcode is moved over a fixed slot scanner, and the information is scanned. However, this technology is slow and can lead to inaccurate reading.

Among these five types of scanners, many retail stores use laser scanners or CCD scanners. The lifespan of a CCD scanner is substantially higher than that of a laser scanner as LED is used in the former case. Laser beams also can cause some health concerns. In addition, CCD scanners are cheaper and have a higher scanning rate than laser scanners. Laser scanners can read barcodes from a longer range. Optical scanners are generally expensive and hence not used widely in the retail stores.

2.2.2.3 *Computer system*

A centralized computer system is used to store the information that is extracted from the barcodes. The scanned images of the barcode are matched with the stored image and the related information is extracted. When the item is sold, the barcode is scanned, and it automatically reduces the stock level. Hence, barcode technology can help to update stock levels automatically in retail stores.

2.3 *RFID technology*

Organizations using RFID technology should understand the basics of how to operate the technology, hence this chapter has focused on the fundamentals of RFID technology. RFID is an emerging radio communication technology that can track and trace items (Asif and Mandviwalla, 2005) in addition to transmitting information without any physical connection between the object and the reader (Tajima, 2007). RFID is an automatic identification system consisting of three major components, namely: (a) RFID tag (also known as transponder), (b) RFID reader (also known as transceiver or interrogator) and (c) RFID controller (also known as RFID host) (Nayak et al., 2007b; Rao et al., 2005; Chao et al., 2007), which are discussed below.

2.3.1 *RFID Tag*

The primary role of an RFID tag is to store data and transmit the data to the receiver. An RFID tag contains the information and can be attached to or embedded into a product, an animal or a person. It consists of two essential components: (a) a tiny semi-conductor chip, where the data is stored and (b) an antenna, to enable the tag to receive and respond to radio frequency from an RFID transceiver (see Figure 2.4). These components are fixed to a rigid substrate and encapsulated with an overlay to make it useful and protect the components. It may also contain a battery as an optional component depending on the type of the tag. The tags contain an Electronic Product Code (EPC) and the information related to the product such as the name of the company, batch and year of manufacturing and product cost (Sarma et al., 2001; Myny et al., 2010).

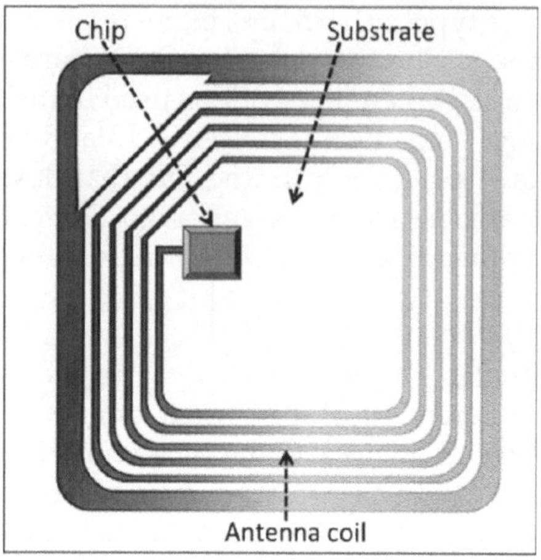

Figure 2.4 An RFID tag with chip and antenna.

The RFID chip is the main source of information storage, which is connected to an integrated circuit (IC) with a microprocessor. The chip can contain varying amounts of information depending on the memory capacity. The antenna (in coil form) in the tag receives the signal from the interrogator or reader, which is processed by the microprocessor. The identifying information stored in each tag is different from the others, which makes it unique.

The other important part of an RFID tag is the antenna, which is responsible for the accuracy of the information delivered to the reader. A good antenna should ensure a proper signal being received by it from the reader and that the right information is being delivered. Improper signal receiving by the antenna may lead to inaccurate and unreliable outcomes. Antenna design depends on the use of the tag and the environment of the tag application. A poor antenna design may lead to misleading and wrong information.

Depending on the source of power supply, RFID tags can be classified as passive, semi-passive and active. Passive RFID tags have no internal power supply. The minute electrical current induced in the antenna by the incoming radio frequency signal provides just sufficient power for the IC in the tag to power up and transmit a response (Bouet and Dos Santos, 2008). Most passive tags are signaled by backscattering the carrier signal from the reader. This means that the antenna must be designed to both collect power from the incoming signal and to transmit the outbound backscatter signal.

The passive tag chip can contain nonvolatile EEPROM (Electrically Erasable Programmable Read-Only Memory) for storing data (Barnett and Liu, 2008). As there is no onboard power supply, the passive tags can be quite small and inexpensive. For example, the commercially available passive tags that can be embedded under the skin are available in the sizes of 0.4 mm × 0.4 mm. Passive tags have practical read distances ranging from about 2 mm up to a few meters, depending on the chosen radio frequency. This is much lower than semi-passive or active RFID tags.

Semi-passive RFID tags are very similar to passive tags except for the addition of a small battery (Zhang and Amin, 2006). This battery allows the IC to be constantly powered and not used for radio signal transmission (Li et al., 2009). This removes the need for the aerial (antenna) to be designed to collect power from the incoming signal. Aerials can therefore be optimized for the backscattering signal. Semi-passive RFID tags are faster in response as compared to passive tags due to the presence of a power source.

Active RFID tags or beacons are so called because they have an internal power source, which is used to power the IC. However, this power is not consumed for transmission of the radio signal (Ni et al., 2004; Jin et al., 2006). The battery is used by the IC while in operation like a TV remote. Hence, the active tags always need to be connected to a power infrastructure or a battery integrated to them. In the former case, the power can last longer depending on the infrastructure, but in the latter case the energy is limited depending on the battery type. Due to the internal power supply, these types of tags can work accurately from a longer range (several hundred feet) even with less powerful readers. The active tags may also have larger memories than the passive tags, as well as the ability to store additional information sent by the transceiver.

Due to the presence of a battery, the active RFID tags are larger in size and complex in design (Bouet and Dos Santos, 2008). For example, at present the smallest active tags are about the size of a coin with 0.5″ diameter. Many active tags have practical ranges of several meters, and a battery life of 3 to 10 years. An airplane transponder is one example of an active RFID tag, which identifies the nationality of the airplane. As passive tags are cheaper to manufacture, have no battery and require no maintenance, most RFID tags in existence are passive tags. Furthermore, passive tags have longer lifespans and are smaller in size to fit perfectly into smaller sized clothing or even the labels in clothing. Therefore, there is a wide acceptance of passive tags for retail purposes.

As mentioned above, passive RFID tags don't need recharging on a regular basis (Vogt, 2002). Therefore, they can be discarded easily without any environmental concerns. Some passive RFID tags can be used as wearable devices by people for monitoring their movement. Examples include: criminals on parole or the military in remote areas. Body parameters such as body temperature and heart rate can also be monitored with passive RFID tags, when additional sensors are fitted into them. See Table 2.2 for a comparison of the three types of RFID tags.

Table 2.2 Comparison between different types of RFID tags

	RFID tag type		
Factors	Active	Semi-passive	Passive
Size	Bigger ______		______ Smaller
Operational range	Up to several meters	Moderate range	Shortest range usually less than 3 meters
Usage	Used on ship containers and railway carriages	Used in pallets, warehouse management and inventory management	Most widely used in retailing, product tracking, material allocation and theft prevention
Required signal strength from reader to tag	Low ______		______ High
Required signal strength from tag to reader	High ______		______ Low
Tag energy source	Battery	Tiny battery; it takes power from RFID reader's radio signal as well	No battery; it takes power from RFID reader's radio signal
Life time	Continuous life up to 10 years	Life up to three years	Short, can be discarded after first use.
Energy usage	Battery used for operation and communication with the RFID reader	Battery power used for standby chip's operation and RFID reader power used for communication	Power derived from RFID reader to operate and communicate
Usage	Higher value products		Lower value products
Price	Most expensive (more than $20 each)	Medium price range ($1 to $7)	Inexpensive (from 20 cents to $5)

Depending on the data storage capacity, RFID tags can be divided into two categories: (a) Read-only (RO) tags and (b) Read-write (RW) tags (Hunt et al., 2007). RO tags have read-only memory, whereas an RW tag's memory is changeable and can store more data depending on the memory capacity. RO tags are like the barcodes, as they are programmed for single use. The information in the RO tags is unchangeable like a CD-ROM (compact disk-read only memory). RO tags possess unique identification codes about the product that can be read by the reader.

Hence, once the product is sold, the tags may become useless. On the other hand, the RW tags can be used multiple times by altering the information contained in the tags. Due to the flexibility of changing information, the RW tags are also called smart tags. The data on an RW tag can be deleted and new data can be reloaded several times like a USB drive. Since an RW tag can be used multiple times, the average cost per use is cheaper than an RO tag, as the RO tag is unusable after one use.

Some RFID tags can have both RO and RW memory simultaneously. This type of tag, for example, can be used in pallet tracking in supply chain. The RO memory can contain the serial number, which is permanent. The RW memory can be varied depending on the contents of the pallet. When the existing products are unloaded, and the new products are loaded in the pallet, the information in the RW memory can be altered using the RFID software system.

In addition to these two major types of RFID tags, there may be some tags known as WORM (write-once-read-many). Like the RO tags, the WORMS can be programmed once with the information from the company but can be used multiple times. If an RO can be compared to a CD-ROM, a WORM can be compared to a CDRW (compact disc-ReWritable). The WORM tags can be programmed with the date of manufacturing or an assembly line number on a production floor.

2.3.1.1 Tag frequency

There are five main frequency bands for RFID tags commonly in use. They are categorized by their radio frequency: low frequency (LF) tags, medium frequency tags, high frequency (HF) tags, ultra high frequency (UHF) tags and microwave tags. RFID tags can be either active, semi-passive (i.e., semi-active) or passive as discussed above (Nayak et al., 2007b; Tesoriero et al., 2008; Rao et al., 2005; Costin et al., 2012). Various applications of RFID tags according to their frequency band are discussed in Table 2.3.

Tags are a key and distinguishing element of RFID systems; they have been transformed into a variety of forms that provide specific characters and functions (Banks et al., 2007). Based on these factors, the cost of each tag varies from 20 cents to 20 dollars. Although barcodes are cheaper than RFID tags, the RFID tags rarely become soiled, dirty or torn and require less human intervention to be read. In addition, tags provide unique traits such as the ability to be read and written from a longer range of communication distances than barcodes.

The read range of a passive tag depends on the frequency of the radio wave. For example, with the low frequency range (100–500 kHz), the passive tags are readable from 4–6 feet (Hunt et al., 2007). This is because of the lack of power source and higher wavelength (λ) of the radio wave. The read range can be increased with the use of higher frequency and the use of active tags.

Table **2.3** RFID frequency bands for various applications

Frequency band	Features	Applications
100–500 kHz (Low)	Short to medium read range; inexpensive; low reading speed	Access control; animal identification; inventory control; car immobilizer
500 kHz–1.2 MHz (Medium)	Medium read range; potentially inexpensive; medium reading speed	Access control; smart cards; library control
1.2–15 MHz (High)	Long read range; high reading speed; line of sight not required; expensive	Railway vehicle monitoring; toll collection systems; pallet and container tracking; vehicle tracking
15 MHz-2.4 GHz (Ultrahigh)	Longer read range; high reading speed; line of sight not required; expensive system	Container and vehicle tracking
≥ 2.4 GHz (Microwave)	The longest read range; high reading speed; very expensive system	Container and vehicle tracking

Source: (Wikipedia, 2018c).

Generally, passive tags are designed to run in low to high frequency ranges, whereas active tags operate in the UHF and microwave frequency ranges. The first RFID design was based on the low to high frequency passive tags.

2.3.1.2 Tag shape

RFID tags may not be always the rectangular- or square-shaped; they can vary, depending on the areas of application. The shapes of RFID tags may be of the following type (see Figure 2.5):

- Rectangular- or square-shaped: These shapes are the most common and found in many retail applications, containers and pallets. Examples: paper tag, EPC tag, ceramic tag and inlay tag.
- Circular-shape: These tags are circular in shape; therefore, the design of the chip and antenna need to fit into that shape. Examples: RFID tags worn by prison inmates, hospital patients or even newborn babies in a hospital; glue tags, button tags and disc tags.
- Key-chain shape: These tags are used in vehicles for automatic toll collection and designed to hold car keys.
- Bullet-shape: These tags are designed to be fitted into animals or birds for data collection on their movement or lifestyle.
- Other-shapes: There are some other shapes of tags designed to fit the area of usage. Examples are ear tags, pocket tags and glass tube tags.

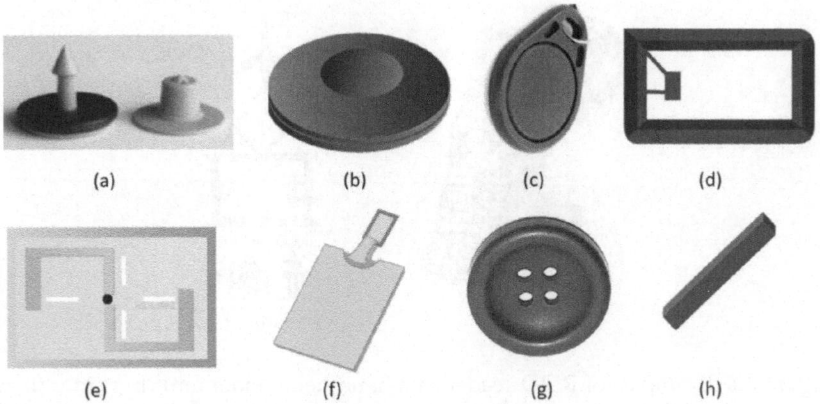

Figure 2.5 Various shapes and types of RFID tags: (a) ear tag, (b) disc tag, (c) key tag, (d) paper tag, (e) ceramic tag, (f) pocket tag, (g) button tag and (h) metal tag.

Source: Mrs. Tran Nha Ca, Lecturer @ RMIT Vietnam.

Some tags can be fitted with additional sensors to detect the temperature and humidity during transportation. Some tags may need to be designed to withstand high temperature, humidity, water and other chemicals. Hence, the external overlay should be made from different materials depending on the application. For use in water, the tags should be water-proof, and for use with chemicals, chemical resistant material should be used. This will increase the price of the RFID tags. Cheap tags or smart tags that are frequently used are prepared with paper or plastic overlays. The overall trend is to minimize the cost of RFID tags while keeping the field of application in mind.

2.3.2 RFID reader

Information from RFID tags is collected by a reader also known as an interrogator (Hunt et al., 2007; Nayak et al., 2007a). The interrogator creates and receives the interrogation signal. The interrogator is a small computer of varying shapes and consists of three components, namely: an antenna, an RF based electronics module and a controller. The antenna is used for signal transmission; the electronics module establishes the communication with the tag; and the controller processes the information. The major functions of an interrogator are: reading the data from the tag; writing the data into the tag; relaying data to and from the computer and providing power to the tag. Various types of RFID interrogators are shown in Figure 2.6.

Modern RFID interrogators are designed to perform additional functions such as: protecting the integrity of data by data encryption;

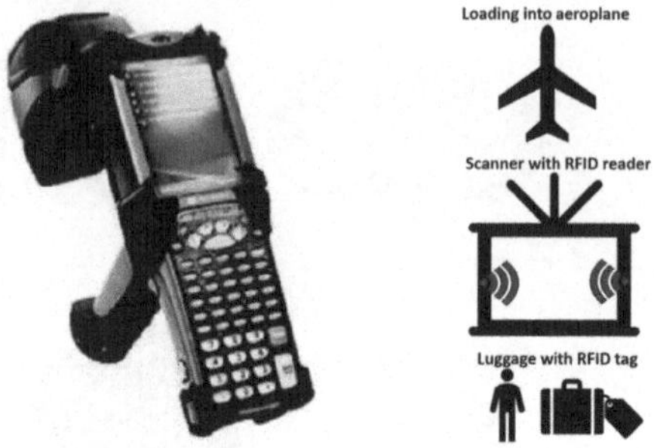

(a) (b)

Figure 2.6 Examples of RFID readers: (a) hand held reader and (b) reader in an airport businesses.

Source: (a) Barburski & Snycerski, 2008.

establishing anti-collision measures to prevent interference between various radio signals; and preventing fraud by authenticating the tags (Hunt et al., 2007). Protecting the data from external attacks is essential to stop hackers trying to steal information such as product details or credit card details used to make the payment. A data encryption and decryption approach is taken to prevent the integrity of data.

Anti-collision measures ensure the interrogator can simultaneously communicate with many tags (Myung et al., 2006). An interrogator sends a signal to receive the information from specific types of products. In a retail store, there may be few products or several hundreds on the shelves. Hence, the information from all the products are collected simultaneously by the interrogator, which can lead to interference of the radio frequencies. The anti-collision measure is a software program or algorithm to avoid the interference between different frequencies.

Advanced algorithms are needed to prevent fraud, which is a common phenomenon in recent times (Rotter, 2008). Hackers will try to manipulate the RFID system and make false payments or modify the information in the tags to pay lower amounts or even disable the RFID tags for shoplifting. Although the additional functions will increase the price of the RFID tags, their use makes the system run more smoothly and the security is enhanced to prevent shrinkage (product loss or theft in a store).

Unlike the barcode system, the RFID system can operate from several meters (Nayak et al., 2007b). This provides flexibility for the placement of the interrogators. In manufacturing facilities, they can be mounted on the walls or placed beneath the working tables. In warehouses the interrogator can be mounted on doors or racks to monitor the movement of products.

Table 2.4 The process of reading active and passive tags

Active tag	Passive tag
An active tag is always ready to receive the signal.	Reader sends radio waves to passive tag to power them.
Reader sends signal to the tag and receives back the information.	Passive tag receives the information and starts transferring the necessary data to the reader.
A battery is always needed for the working of active tags.	It can last longer, as no battery is needed for its working.
The lifetime is based on the battery life.	Generally discarded after one use.
The larger size and cost limits their application.	As they are smaller in size and cheaper, they are widely used in various sectors.

They can also be mounted to moving tools such as conveyors, forklifts or lifting equipment. There is a subtle difference between data extraction techniques for an active tag and a passive tag, which are shown in Table 2.4.

The reader is also connected to an antenna or multiple antennas. Within the range, a reader can read all the tags quickly and sequentially by using various techniques. Those techniques are grouped as "singulation," working by allowing only specific serial numbers of a tag to respond. On the other hand, RW tag data can be changed and edited in real time by certain types of readers or reader/writer devices. Hence, these types of devices are very useful in including additional information, due to their dual nature.

2.3.3 RFID controller

The function of the RFID controller is to integrate and manage the whole RFID system. Hence, the controller is also known as the brain of the RFID system like the central processing unit (CPU) in a computer. The controller consists of a miniature computer or a workstation with software (middleware) installed in it, or it can be a network of several workstations. The controller can use three levels of middleware that are applied based on industry context and requirements (Assif & Mandiwalla 2005):

- The first level is software that deals with the industries' connectivity.
- The second level is the management software that is used to create and monitor various applications in an industry.
- The third level is the intermedium software to connect applications and devices, for instance: RFID interrogators and shop-floor machines.
- In addition, the controller also has a data management system and an enterprise application interface.

The RFID controller has several functions including: tracking the location of the objects in a retail store and tracking inventory. When a product is sold, or a new product is received, inventory is automatically updated by the controller. Similarly, when the product on a shelf reaches a minimum threshold, it sends an alert to the store manager.

2.3.4 *RFID system in operation*

Figure 2.7 illustrates how an RFID system works. When the information is needed from an RFID tag, a command is given from the controller (or CPU) to the reader. The reader sends the information and communicates with the RFID tag (which contains digital information) through its antenna. A collection of wired and wireless networks constitutes the communication infrastructure that carries out a series of information transfer actions to deliver the data stored in a tag to the reader. Readers are just like the barcode sensors, which broadcast a radio signal through the antenna to the tag. The tag then responds to the radio wave, and the data can be read from the chip of the tag by the reader.

The reader can read data from multiple tags simultaneously. Readers also decode the signal provided by the tags and transmit to the CPU. Suitable software receives and interprets the information collected from the tags and stores it. Simultaneously multiple data can be collected and stored with a high degree of accuracy compared to the existing barcode systems using one tag per scan. As mentioned earlier, passive tags collect energy from a nearby RFID reader's interrogating radio waves, whereas the active tags have a local power source (such as a battery) and may operate from several meters from the RFID reader. Unlike a barcode, the tag need not be within the line of sight of the reader, so it may be embedded in the tracked object. RFID is one method for Automatic Identification and Data Capture (AIDC), which is different from the barcode technology.

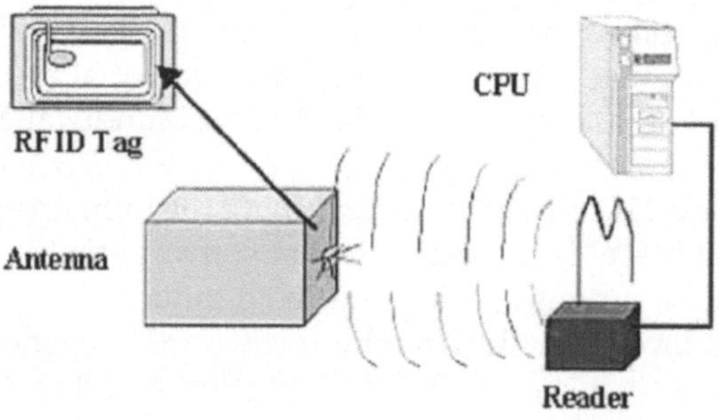

Figure 2.7 RFID system in operation.

2.4 RFID vs barcode system (technical differences, advantages and disadvantages)

Both barcode and RFID technologies are used for product tracing and tracking. In addition to these two technologies, other tools that are used for product tracing and tracking are EPC and web portal tools (Staake et al., 2005). The RFID system and EPC are technologies for the future and are used in tandem. RFID systems enhance the unique identification of each individual product, case or pallet in the supply chain process. Each product must be tagged with an RFID tag programmed with unique EPC and stored in memory. The signals from RFID tags can be collected by the readers and transmitted to the information collection center. On the other hand, barcodes are also used to uniquely identify individual products or pallets. Barcode systems, although used for product information and inventory control, have several drawbacks as compared to the RFID systems (Delen et al., 2007; Kärkkäinen, 2003).

RFID tags can be used in manufacturing to identify product code, manufacturer's identity, manufacturing date, pallet number and warranty period, something that may not be possible with a barcode (Hunt et al., 2007). Furthermore, if a product needs to be transported in a special condition such as freezing temperature, the temperature sensors in the RFID tag can store that information during the transportation process, which is not feasible with barcode technology. Table 2.5 provides a comparison of barcode technology and RFID technology.

A detailed difference between the barcode technology and RFID technology has been given in this section.

2.4.1 Limited information

The amount of information stored in a barcode is very limited as compared to RFID tags (Bolic et al., 2010). The smallest barcodes (1D) can hold up to eight numeric characters (i.e., UPC E symbols), which is just a few bytes of data. However, 2D barcodes can contain up to 1000 ASCII characters or 100 bytes of data. On the other hand, an RFID tag can store much more data. The smallest tag can hold 128 kilobytes of data, which is much higher than the information stored in barcodes. Larger RFID tags can hold up to several megabytes of data, which has increased with technological improvements. In addition, the barcode system is less accurate as compared to the RFID system when multiple sources of data are used.

2.4.2 Human interaction

A barcode needs human interaction for getting the product information. It requires line-of-sight access to an optical scanner for extracting the

Table **2.5** Comparison of RFID technology with barcode technology

Parameter	RFID technology	Barcode technology
Principle of operation	The use of radio signal and electromagnetic field for data collection	Optical reading of printed codes using laser for data collection
Materials	Plastic, metal and semiconductors	Paper
Amount of information	Up to some megabytes (depending on the type)	Equivalent to 100 bytes
Range of operation	Several metres to kilometres (system dependent)	Several centrimeters or line-of-sight
Reading rate	Up to 1000 tags simultaneously	Only one at a time
Technology	Contactless and automatic read by the reader	Manual and need to be presented at the scanner
Principle of operation	Radio frequency	Optical (Laser)
Reader location	Anywhere in the range	Line-of-sight
Reusability	Yes	No
Read/write	They are read/write type or read-only type	Read-only type
Automatic operation	Can work automatically without human interaction	Need human interaction for reading
Security	High	Low
Cost	High	Low
Identification	Uniquely identify each object	Can identify the product type, but not uniquely
Signal obstruction	Can interfere with metals and solids, leading to wrong information	Damaged, dirt covered, or erased barcode can't be read

information. Hence, without human interaction data can't be extracted from the barcodes. Therefore, automatic reconciliation of stock is not possible in a barcode system. On the other hand, RFID tags can be read by readers without human intervention from several centimeters to several meters. Hence, stock can be reconciled easily with the RFID tags whenever needed without any human interaction.

2.4.3 *Reusability*

Barcodes need to be replaced if the information it contains needs modification. A new printed barcode should be prepared to replace the old ones. Once the product is sold, barcodes cannot be reused. However, in the case of RFID tags the information can be modified at any stages of the supply chain by the interaction between the microchip and the reader software, depending on the type of RFID tag.

2.4.4 Security issues

An RFID tag is specific to each product, whereas a barcode is not. However, both barcode and RFID technologies are not secured, as the information can be modified by attackers. For example, although RFID systems can improve productivity and efficiency, they can also expose an organization to new threats to security and privacy (Phillips et al., 2005; Bogdanov et al., 2008; Ayoade, 2007). The information in the RFID tags can be accessed by hackers. Similarly, attackers can easily replace or modify the original barcode tags with duplicate tags to provide wrong information.

2.4.5 Cost

Barcode tags are much cheaper than RFID tags. A printed barcode can be as cheap as 1 cent, whereas an RFID tag can cost in the range of 20 cents to $1 or even higher. However, the reusability of RFID tags can bring the cost down. The cost of RFID tags depends on the amount of data stored and the readability range. The more an RFID tag is used, the cheaper its price. Furthermore, when the benefits are considered, the cost of the tags should not be taken into consideration. The benefits achieved from RFID tags compensate for the high cost.

2.4.6 Reliability and range

The reliability of barcodes plays a significant role in the supply chain. Barcodes that are damaged or badly folded are not easily read by a scanner. In those cases, the information needs to be typed manually, involving the process of locating information such as the price, all of which is time-consuming. RFID tags are free from this problem and more efficient in delivering the data when needed.

Many barcode readers can read a tag when it is only few centimeters away or within the line-of-sight. Beyond that, the readers can't read. However, some specially designed barcode readers with advanced technology can even read from some meters. RFID tags can be read from several meters without the product being removed from the shelves. RFID technology is more reliable than barcode technology; they are more accurate and hard to make physically unreadable.

2.5 Conclusion

RFID is an emerging technology for fashion and textile supply chain and manufacturing. Barcode technology existed for several years and is currently in use by several fashion brands globally. However, RFID technology is being increasingly used in fashion supply chain from fiber production to retail stores. The ability of RFID technology to trace objects, identify the object, provide real-time information and accuracy of information are the major reasons for the wide acceptance of the technology, which is not feasible with barcode technology. All the role players in the fashion supply chain starting from fiber producers, yarn manufacturers, fabric and garment manufacturers can track the information shared and know the status of the merchandise at any time, which a barcode system cannot provide.

Many retail stores use both RFID and barcode technology side-by-side. Some global fashion brands have made RFID technology mandatory in the manufacturing industries and retail stores or even with their suppliers. Many other brands are switching to RFID technology after realizing its tangible benefits. However, there are many limitations that need to be overcome for RFID to be successfully adopted by all the fashion and textile industries around the world.

References

Airport-Business. 2009. *Making RFID work – the most effective solution to lost baggage?* [Online]. Available: http://www.airport-business.com/2009/03/making-rfid-work-the-most-effective-solution-to-lost-baggage/ [Accessed 15 May 2018].

Akeroyd, A. 2010. The Barcode Revolution. *Journal of the Society of Archivists*, 31, 51-62.

Asif, Z. & Mandviwalla, M. 2005. Integrating the supply chain with RFID: A technical and business analysis. *Communications of the Association for Information Systems*, 15, 24.

Ayoade, J. 2007. Roadmap to solving security and privacy concerns in RFID systems. *Computer Law & Security Review*, 23, 555-561.

Banks, J., Pachano, M. A., Hanny, D. & Thompson, L. G. 2007. *RFID Applied*, John Wiley & Sons.

Barcodeart. n.d. *Barcode Diagram* [Online]. Available: http://www.barcodeart.com/artwork/collages/diagram/index.html [Accessed 18 August 2018].

Barnett, R. E. & Liu, J. 2008. An EEPROM programming controller for passive UHF RFID transponders with gated clock regulation loop and current surge control. *IEEE Journal of, Solid-State Circuits*, 43, 1808-1815.

Basker, E. 2012. Raising the barcode scanner: Technology and productivity in the retail sector. *American Economic Journal: Applied Economics*, 4, 1-27.

Bogdanov, A., Leander, G., Paar, C., Poschmann, A., Robshaw, M. J. & Seurin, Y. 2008. Hash functions and RFID tags: Mind the gap. *Cryptographic Hardware and Embedded Systems–CHES 2008*. Springer.

Bolic, M., Simplot-Ryl, D. & Stojmenovic, I. 2010. *RFID Systems: Research Trends and Challenges*, John Wiley & Sons.

Bouet, M. & Dos Santos, A. L. RFID tags: Positioning principles and localization techniques. Wireless Days, 2008. WD'08. 1st IFIP, 2008. IEEE, 1-5.

Chao, C.-C., Yang, J.-M. & Jen, W.-Y. 2007. Determining technology trends and forecasts of RFID by a historical review and bibliometric analysis from 1991 to 2005. *Technovation*, 27, 268-279.

Costin, A., Pradhananga, N. & Teizer, J. 2012. Leveraging passive RFID technology for construction resource field mobility and status monitoring in a high-rise renovation project. *Automation in Construction*, 24, 1-15.

Delen, D., Hardgrave, B. C. & Sharda, R. 2007. RFID for better supply-chain management through enhanced information visibility. *Production and Operations Management*, 16, 613-624.

Hong-Ying, S. The application of barcode technology in logistics and warehouse management. Education Technology and Computer Science, 2009. ETCS'09. First International Workshop on, 2009. IEEE, 732-735.

Hunt, V. D., Puglia, A. & Puglia, M. 2007. *RFID: A Guide to Radio Frequency Identification*, John Wiley & Sons.

Jin, G.-Y., Lu, X.-Y. & Park, M.-S. An indoor localization mechanism using active RFID tag. Sensor Networks, Ubiquitous, and Trustworthy Computing, 2006. IEEE International Conference on, 2006. IEEE, 4 pp.

Kärkkäinen, M. 2003. Increasing efficiency in the supply chain for short shelf life goods using RFID tagging. *International Journal of Retail & Distribution Management*, 31, 529-536.

Katona, M. & Nyúl, L. G. Efficient 1D and 2D barcode detection using mathematical morphology. International Symposium on Mathematical Morphology and Its Applications to Signal and Image Processing, 2013. Springer, 464-475.

Kim C., Kim K., Choi Y. C. 2012. The Framework Development of a RFID-Based Baggage Handling System for Airports. In: Yeo S. S., Pan Y., Lee Y., Chang H. (eds) Computer Science and its Applications. Springer, Dordrecht.

Kippelen, B. & Durgin, G. D. 2010. Thin flexible radio frequency identification tags and subsystems thereof. US Patent: US7642918B2.

Li, X., Zhang, Y. & Aamin, M. G. Multifrequency-based range estimation of RFID tags. IEEE International Conference on RFID, 2009. IEEE, 147-154.

Longacre, A. 2004. Method for processing images captured with bar code reader having area image sensor. US Patent: US20030197063A1.

Myny, K., Steudel, S., Smout, S., Vicca, P., Furthner, F., Van Der Putten, B., Tripathi, A. K., Gelinck, G. H., Genoe, J. & Dehaene, W. 2010. Organic RFID transponder chip with data rate compatible with electronic product coding. *Organic Electronics*, 11, 1176-1179.

Myung, J., Lee, W. & Srivastava, J. 2006. Adaptive binary splitting for efficient RFID tag anti-collision. *IEEE Communications Letters*, 10, 144-146.

Nadeem, M. & Velasco, T. 1993. Knowledge-based approach to quality control assurance using bar code identification systems. *Computers & Industrial Engineering*, 25, 41-44.

Nayak, R., Chatterjee, K., Khandual, A. & Gupta, T. 2007a. Scope & limitations of RFID. *Indian Textile Journal*, 117, 25-29.

Nayak, R., Chatterjee, K., Khurana, G. & Khandual, A. 2007b. RFID: Tagging the new era. *Man-Made Textiles in India*, 50, 174-177.

Nayak, R. & Padhye, R. 2017. *Automation in Garment Manufacturing*, Woodhead Publishing.

Nayak, R. & Padhye, R. 2018. Introduction to automation in garment manufacturing. In: Nayak, R. & Padhye, R. (eds.) *Automation in Garment Manufacturing*. Elsevier.

Nayak, R., Singh, A., Padhye, R. & Wang, L. 2015. RFID in textile and clothing manufacturing: technology and challenges. *Fashion and Textiles*, 2, 1-16.

Ni, L. M., Liu, Y., Lau, Y. C. & Patil, A. P. 2004. LANDMARC: indoor location sensing using active RFID. *Wireless Networks*, 10, 701-710.

Phillips, T., Karygiannis, T. & Kuhn, R. 2005. Security standards for the RFID market. *Security & Privacy, IEEE*, 3, 85-89.

Rao, K. S., Nikitin, P. V. & Lam, S. F. 2005. Antenna design for UHF RFID tags: A review and a practical application. *IEEE Transactions on Antennas and Propagation*, 53, 3870-3876.

Rotter, P. 2008. A framework for assessing RFID system security and privacy risks. *IEEE Pervasive Computing*, 70-77.

Sarma, S., Brock, D. & Eengels, D. 2001. Radio frequency identification and the electronic product code. *IEEE Micro*, 21, 50-54.

Sriram, T., Rao, K. V., Biswas, S. & Ahmed, B. Applications of barcode technology in automated storage and retrieval systems. Proceedings of the 22nd International Conference on Industrial Electronics, Control, and Instrumentation, 1996. IEEE, 641-646.

Staake, T., Thiesse, F. & Fleisch, E. Extending the EPC network: the potential of RFID in anti-counterfeiting. Proceedings of the ACM symposium on Applied Computing, 2005. ACM, 1607-1612.

Tajima, M. 2007. Strategic value of RFID in supply chain management. *Journal of Purchasing and Supply Management*, 13, 261-273.

Tesoriero, R., Gallud, J., Lozano, M. & Penichet, V. 2008. Using active and passive RFID technology to support indoor location-aware systems. *IEEE Transactions on Consumer Electronics*, 54, 578-583.

Vogt, H. Efficient object identification with passive RFID tags. International Conference on Pervasive Computing, 2002. Springer, 98-113.

Wikipedia. 2018a. *Barcode* [Online]. Available: https://en.wikipedia.org/wiki/Barcode [Accessed 10 July 2018].

Wikipedia. 2018b. *PDF417* [Online]. Available: https://en.wikipedia.org/wiki/PDF417 [Accessed 17 August 2018].

Wikipedia. 2018c. *Radio-frequency identification* [Online]. Available: https://en.wikipedia.org/wiki/Radio-frequency_identification [Accessed 25 July 2018].

Yaqoob, Z. & Riza, N. A. 2004. Passive optics no-moving-parts barcode scanners. *IEEE Photonics Technology Letters*, 16, 954-956.

Zhang, Y. & Amin, M. G. Localization and tracking of passive RFID tags. Defense and Security Symposium, 2006. International Society for Optics and Photonics, 6248, 1-11.

chapter three

Application of RFID in fiber production and yarn manufacturing

KEYWORDS: *fiber production, yarn manufacturing, RFID technology, bobbin management, yarn quality tracking*

3.1 Introduction

The manufacturing of garments has shifted from developed countries to developing countries such as Bangladesh, Vietnam, Laos, Cambodia and Indonesia (Gereffi and Frederick, 2010). The major reason is the low labor cost in developing countries, which helps to keep the final product cost low (Nayak and Padhye, 2015:, Nayak and Padhye, 2014). The garment manufacturing industries in developing countries rely on the fabrics, threads, trims and accessories' suppliers located in the upstream of the supply chain process. Similarly, fabric manufacturers rely on yarn manufacturers and yarn manufacturers rely on the fiber producers. The supply chain process of the garment manufacturing industry is shown in Figure 3.1, where each downstream player relies on the upstream player.

In the recent competitive supply chain dynamics, the role of the upstream players is not just supplying the raw materials to the downstream organizations. The successful operation involves maintaining a strategic relationship among all the organizations, horizontally and vertically (Christopher, 2016). Several times, the inaccuracy in the information flow and unavailability of real-time information creates a hindrance to a good relationship (Thomassey, 2010). Therefore, it is essential for organizations to provide accurate and timely information. The use of Radio Frequency Identification (RFID) technology can help to achieve these objectives in the supply chain process (Nayak et al., 2015; Nayak et al., 2007a).

With the phase out of multi-fiber arrangement (MFA) in 2005, the textile and garment industries have the opportunity of expanding their business to boost exports (Lan and Zhu, 2014). This has led to intensive competition between the garment industries in many developing countries. Therefore, the garment industries are seeking alternative routes to gain competitive advantage. One approach is the response time or lead time to a buyer's need

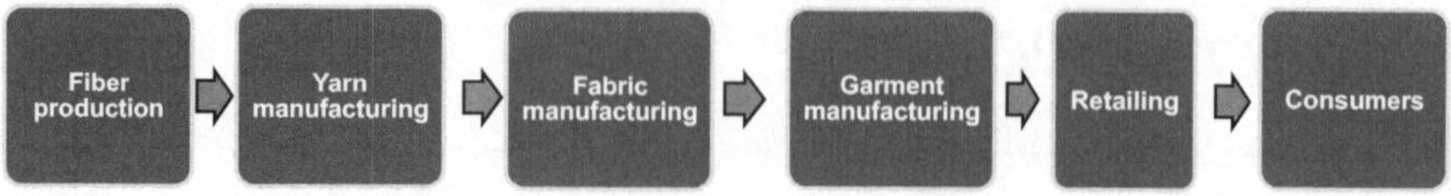

Figure 3.1 Process sequence for garment manufacturing.

(Kadarusman and Nadvi, 2013). Hence, apparel manufacturers are trying to reduce their lead time in addition to providing innovative products and maintaining consistency in the cost and quality.

Various approaches such as automation and implementation of new technologies can help in reducing the lead time (Nayak and Padhye, 2017; Nayak and Padhye, 2018). Textile industries are heading towards automation in many of their departments to achieve higher productivity and efficiency (Joshi and Singh, 2010). Furthermore, the application of RFID technology is also becoming popular for better information management and process control, which can help to bring down the lead time by providing accurate and real-time information (Nayak et al., 2007b).

This chapter focuses on the use of RFID technology in the textile manufacturing process, from fiber production to yarn manufacturing. In fiber production RFID tags can be used to monitor individual plant and animals; and in spinning industries, the use of RFID technology can help in managing the bobbins, cones, material management and warehouse management, which are also discussed in this chapter. Furthermore, the use of RFID technology in machine maintenance and yarn quality management has also been discussed in this chapter.

3.2 *RFID in fiber production*

Fibers are the basic raw materials for manufacturing garments. Textile fibers are generally derived from natural resources and petroleum byproducts. Depending on the resources, the fibers can be classified into two groups: natural fibers and synthetic fibers (Gupta and Kothari, 2012). Natural fibers are derived mainly from plants (such as cotton, flax, kapok, sisal and jute) and animals (such as wool and silk). On the other hand, synthetics fibers (such as nylon, acetate, acrylic and polyester) are derived from petroleum resources. Synthetic fibers such as viscose, bamboo and lyocell can also be prepared from natural resources by using artifically-made processes, which are known as regenerated fibers.

3.2.1 Plant fiber production

Production of natural fibers such as cotton, hemp and flax mainly depend on the farming activities of fiber growers. As such, the scope of application of RFID technology during the plant fiber production has not received much focus. This can be due to the nature of farming, remote locations and the expensive process of implementing RFID technology. However, some fiber growers are attempting to implement RFID technology in various plant fiber productions (Ron and Greg, 2016). Similarly, there are several areas of application of RFID technology during animal fiber production such as wool (Brown et al., 2015) and silk (Nivaashini et al., 2018), which are discussed in the subsequent section.

RFID technology can be used for tracking of individual plants, like tracking of an individual animal (Saraswat and Robbins, 2011). Specially designed RFID tags can be fitted into the stem of the fiber plant. The tags can contain information on the variety of the fiber, plant growth record and watering conditions. Furthermore, the seed record, such as seeds that are selected from plants that grow faster and free from disease, can be stored. RFID tags can improve the accuracy of keeping records on individual plant data, which can be used to select good quality fiber from an early stage. An RFID tag used in the tracking of fiber plants is shown in Figure 3.2.

RFID tags can also be used to store information on the type of chemicals, quality of chemicals, type and amount of fertilizers and pesticides applied to the plant. This may be helpful to know the conditions the fibers are subjected during farming. Similarly, the concept of organic farming can be verified from the information stored on the RFID tags. RFID tags can also provide information to the consumers on the history of chemicals used to grow the plants. Hence, the consumers can make appropriate decisions on sustainability when buying clothes of these fibers.

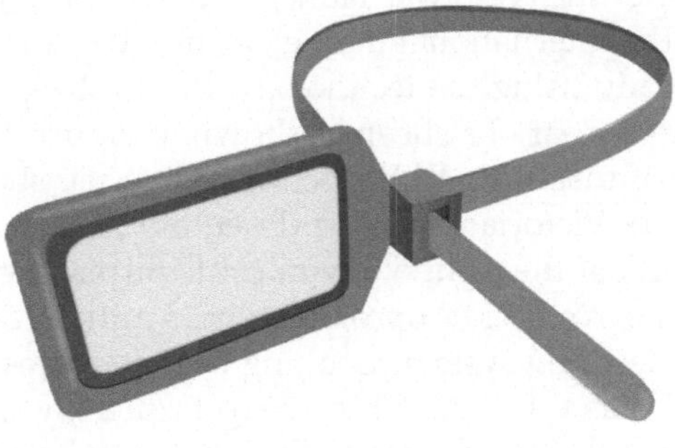

Figure 3.2 RFID tags used for identifying plants.

Source: Mrs Tran Nha Ca, Lecturer @ RMIT Vietnam.

3.2.2 Animal fiber production

In this section, the application of RFID technology in animal fiber harvesting is discussed.

3.2.2.1 Wool

Sheep are grown on farms for meat, milk and wool fiber. The use of RFID technology in farms raising sheep for exporting the meat has been well documented (Age, 2005). However, there is no strong evidence of the use of RFID technology on sheep farms engaged only in wool fiber harvesting. Therefore, the author visited three sheep farms in Australia and contacted three Indian sheep farms to explore the status of application of RFID technology.

Upon his investigation of the Australian sheep farms, the author established that the use of RFID technology is very limited. Two of the farms did not install the technology due to additional cost of implementing the RFID system. However, the third farm, although it had implemented the technology due to government pressure, has faced several problems while working with the RFID system. The major problem is the use of low frequency (LF) RFID tags, which can only be scanned from a few centimeters and the scan rate is very slow. Hence, managing a large flock of sheep using the LF tags is a challenging task. The Indian farms on the other hand mentioned that they are not familiar with the RFID technology. The reason was attributed to the remoteness of the farms and a lack of government initiatives to mandate the application of the RFID technology.

Although, the idea of implementing RFID identification technology in cattle industries in Australia was coined in 1999 by the Department of Primary Industries (DPI), it was not very successful (Friedlos, 2008). In 2003, the National Livestock Identification System (NLIS) was launched, which was the largest RFID application in Australia (NLIS, 2018). The choice of adopting the technology was left up to the farmers to select it voluntarily if it made sense for them commercially. Some farms adopted the technology after the NLIS was launched and some adopted it to meet the demands of the international market, as many farms in the USA and Europe were already using the technology. In the sheep farms the RFID tag is pierced into the ear of a sheep as shown in Figure 3.3.

The benefits of installing RFID technology to tag sheep was investigated by DPI in six Victorian suburbs (Friedlos, 2008). It was found that the implementation of the technology in cattle farms improved the accuracy of tracking the livestock. Approximately 25 million cattle movements were traced by the RFID system. Looking at that success, DPI provided subsidized RFID tags ($1.35 AUD) to sheep farmers, with the hardware and software at a subsidized rate, so that they can be successful (Dave, 2008). RFID technology can help to monitor an individual sheep's health, weight gain or loss and any special diseases. This can help to identify a sheep's dietary requirements, shearing date and health.

(a) (b)

Figure 3.3 RFID tag used in sheep farming: (a) RFID tag and (b) RFID tag pierced in a sheep's ear.

Sources: (a) (Maselyne, 2014) and (b) (Wikipedia – John Haslam, 2008b).

The use of RFID technology can help to identify the sheep with the best and worst quality wool fiber. Hence, special care can be taken to improve the wool quality of the sheep with poorer wool fibers, and the best quality sheep can be managed appropriately to maximize profit. Similarly, a ewe can be tagged with an RFID tag to find the number of hours it spends with its mother (Brown et al., 2015). This indicates the amount of feeding from the mother, and whether it is enough or not so that dietary requirements can be changed accordingly. RFID readers are being installed at the gates, passageways, shearing sheds and narrow races in the sheep farm to read the tags.

As per the guidelines of the NLIS and the decision of the Australian Wool Exchange (AWEX), many wool farms have selected LF RFID tags (Rain-RFID, 2017). However, due to the low reading range of LF tags, AWEX has recently suggested a switch to ultra high frequency (UHF) tags, as hundreds of tags can be scanned simultaneously from 6–8 meters (Annabelle, 2017a). The UHF tags can provide a high level of reading speed and can work in harsh conditions. The replacement cost will require a large investment from the exchequer (Australian federal budget). However, the return on investing in UHF tags should offset the rising labor costs seen in developed countries such as Australia.

RFID identification technology in animal fiber farming is also gaining popularity in China and some parts of India, which are the leading wool manufacturers in the world. Many farms, especially in China, are switching from traditional methods of identifying sheep, such as ear tags (nickel, copper or 2D barcode tags), earmarking (notching or punching), hair dye, toe clipping, implanting microchips, manual tattoos (tail, foot pad or ears) and automated tail tattoos to RFID technology. The information on feeding, yield, breeding and health of sheep can be stored using RFID technology.

The RFID tags used for sheep tracking are specially designed with high quality material to fit the ear of the sheep (see Figure 3.3). The tags

Table 3.1 Specification of RFID tags used for animal tracking

Specification	Specification (Tag 1)	Specification (Tag 2)
Tag type	Ear tag	Bio glass tag
Frequency	134 kHz	125 & 134 kHz
Substrate	TPU (thermoplastic polyurethane)	3M glass + adhesive
Operating temperature	20 to 100° C	−20 to 65° C
Application	Sheep, horse and birds	Animal tracking and inventory management

track the exact location of the sheep when needed and maintain the database effectively. Any attempt to remove a tag breaks the tag and connectivity is lost. This helps insurance companies to trace the sheep accurately. The specifications of two RFID ear-tags used for animal tracking is given in Table 3.1 (Shashikant, 2017).

Although tracking sheep and other animals has been enforced in many European countries such as France and Germany, the farms were reluctant to adopt the technology. Some of the farms moved to court against the rules demanding not to adopt the RFID technology due to high cost. There are still some pending cases on the court on the adoption of RFID technology by sheep farms.

3.2.2.2 Silk sericulture

The silk fiber is produced from the larva of silkworms undergoing complete metamorphosis. The process of silk farming is known as "sericulture," which is practiced in many countries such as India, China, Japan, Russia, Brazil and France. The silkworm develops from the egg stage and passes through the larvae and cocoon stage and finally converts to the adult moth, which cuts open the cocoon and flies away. The adult moth lays eggs and the process continues. The life cycle is completed in specific types of plants such as the Mulberry. Temperature, humidity, feeding and sterilization are some of the important parameters affecting the growth of the silkworm. Hence, keeping a record can help in identifying good quality silk fibers, which can be managed by the RFID technology.

Monitoring the silkworm's lifecycle can help to control the quality of the silk. The use of Internet of Things (IoT) is growing in the use of activation, detection and transmission of information in farming (Nivaashini et al., 2018). In this regard, RFID and Bluetooth technology are becoming increasingly useful. Hence, the combination of IoT and RFID technology can be jointly used to collect information on silkworm growth. Some industries have already implemented RFID technology in silk sericulture (Nivaashini et al., 2018).

3.2.3 Fiber bale tracking

Many natural and synthetic fibers are transported in the form of large rectangular packages known as bales. A bale can contain 150–230 kg of fiber, rigidly compressed and packed in plastic bags. Tracking of woolen bales has been in practice in some countries. For example, Western Australia's wool clip is commercially exploring the viability of implementing RFID in wool bale transportation (Annabelle, 2017b). In 2018, 150,000 bales will be tracked from farm to shipping point. The tags will cost about $1 and can be scanned using mobile phones. This initiative is started by AWEX to gain the commercial benefits of traceability, e-Bale's logistic and provenance opportunities.

Like wool bale tracking, RFID technology can also be used to track cotton fiber bales. After plucking the cotton fibers, they are subjected to ginning process to remove the husks and leaves from the fiber. Then the fibers are tightly packed into compact shapes called bales for easy transportation. The ginners need to confirm the correct bale is identified and loaded in to the truck before they leave to the buyers. The current practice to identify and track the bales during ginning in many countries is by manual means using printed barcodes or other numbers attached to the bale, which is cumbersome and time consuming.

Some countries are now switching to the use of RFID technology, to explore the potential benefits of RFID technology. The RFID tags containing the information on the farmer, type of cotton, weight and place of production are attached to the bale. As they are loaded into the delivery truck, the database is automatically updated as the reader reads the information. An example of RFID includes Texas Cotton Ginners working in Australia to develop RFID tags for the cotton bales in the ginning facilities (Ron and Greg, 2016). This technology prevents the wrong cotton being transported to the buyer or it identifies the missing modules in the ginning process (Claire, 2017). The use of RFID technology in ginning farm in Australia is shown in Figure 3.4. Similarly, RFID technology can be implemented during the transportation of the bales of synthetic fibers including the manufacturing process.

3.2.4 Synthetic fiber manufacturing

Synthetic fibers are produced by various processes such as melt spinning, dry spinning and wet spinning. In the fiber manufacturing industries there are several machines used to make the fibers, followed by drawing, heat setting, texturing and tow-to-staple cutting. Synthetic fibers such as polyester, nylon and acrylic need various types of chemicals for the fiber synthesis. The RFID technology can help to keep the stock of all the materials and provide real-time information during fiber manufacturing.

Figure 3.4 A built-in RFID reader to link each unit of harvested cotton to its farmer, field, date, time, latitude and longitude.

Source: ShutterStock.

The types of machines, maintenance scheme of machines and other machine setting parameters can be managed using the RFID system.

For example, during the polyester fiber formation, the right grade of purified terephthalic acid (PTA) and monoethylene glycol (MEG) need to be mixed (Gupta and Kothari, 2012). If any other substances are selected by mistake, the alarming system will confirm the wrong materials are selected. The information on the extrusion temperature, speed of extrusion, drafting speed and winding speed can also be saved in the RFID tags. This information can be used for referencing when needed in the assembly line. The RFID tags can be used to update the information relating to the texturing process such as type of texturing, temperature and air speed.

The fiber extrusion machines, texturing machines and other machines need regular maintenance. It is very hard to keep track of all the maintenance done or scheduled in the future. Hence, the RFID technology can help to keep the maintenance record or even the record of the spare parts needed by various machines. Regular alerts can be received from the RFID system if the maintenance schedule has not been maintained or the components are running out.

3.2.5 *Managing fiber testing*

Various natural and synthetic fibers are tested for their fineness, tenacity, elongation and uniformity. Many of the fibers resemble each other in appearance and hard to be distinguished. In the fiber testing centers, large numbers of fibers are tested on a daily basis. For example, the Australian Wool Testing Authority (AWTA) tested 2,006,132 bales in the financial year 2016/2017 (AWTA, 2018). Managing such a large number of samples is a difficult task and often leads to inaccuracies in information keeping and sampling process. The use of RFID technology can help to improve the accuracy of information and storing the test results of individual fibers. Each fiber coming for testing needs to be fixed with an RFID tag when going to the sampling department. All the test parameters, fiber grading and supplier's details can be stored by the RFID tags easily, which can also be used for costing purposes.

3.3 *Yarn manufacturing*

Yarn manufacturing or spinning is the process of converting fibers in to yarns. Different types of fibers such as cotton, wool, flax, hemp, nylon and polyester are spun as single fibers or as blends of multiple fibers in different spinning systems such as ring, rotor and airjet to produce yarn. The ring spinning system is widely used due to superior yarn qualities and its versatility to spin a wide range of yarn counts. However, the problem of ring spinning is low productivity and the high number of machines required in the process using substantially higher electricity. Hence, new spinning systems such as rotor and airjet came into play where the strength of the yarn is not a prime factor. The new spinning systems reduce the carbon footprint to a significant amount (van der Velden et al., 2014). The sequence of spinning operations is shown in Figure 3.5.

In the spinning industry, there are several types of raw materials such as fibers that vary in their composition, color, fineness, strength and uniformity. In addition, the spinning industries must deal with a range of materials used for the maintenance of the machines starting from the blow room to winding. Almost all the materials in the spinning industries are dealt manually, which is a time-consuming process and inaccurate. Hence there is a good scope within RFID technology to be implemented in the identification of materials, keeping record of the production, quality management, real-time information availability and process control.

3.3.1 *Bobbin management*

The spinning process can be of two types: staple or continuous filament yarn spinning. The staple yarn production can be applied to both natural and

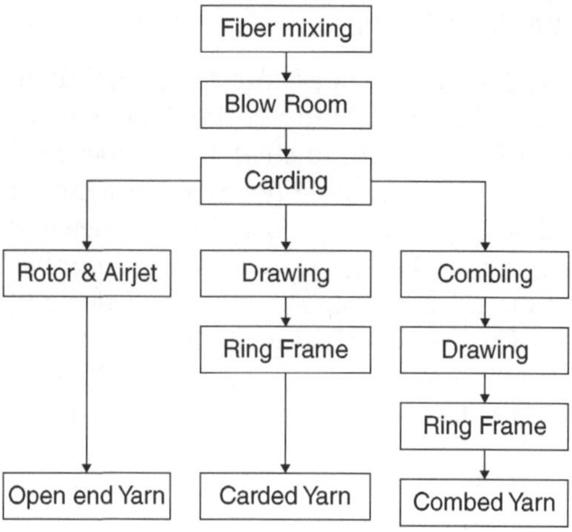

Figure 3.5 The process sequence in yarn manufacturing.

synthetic fibers whereas the filament yarns are produced for synthetic fibers only. Production of staple yarns needs several different machines compared to the filament yarns. Staple yarn spinning needs blow room, carding and draw frame (for open end spinning such as rotor and airjet); roving frame and ring spinning are needed for producing ring yarns. Filament yarn production requires fiber/filament extrusion machines and texturing machines.

The spinning processes for staple spinning involves ring spinning, rotor spinning and air-jet spinning. Among these, ring spinning is the traditional process of yarn manufacturing that has existed since 1779. Ring spinning is the most versatile process, as it can produce a wide range of yarns with the highest strength (Shaikh and Bhattacharya, 2016). Hence, it is the most widely used spinning system to produce yarns. In ring spinning the yarns are collected in the bobbins, which are mounted over spindles. The spindles are regularly maintained to produce consistent yarn quality. However, some of the spindles malfunction despite regular maintenance. The management of a bobbin to spindle allocation can help to produce better yarn quality.

The number of spindles in a modern ring frame can be up to 1000, which makes it difficult to keep a track of the bobbins, producing defective yarns. Although some industries such as the WUXI Textile Mill No. 1, Jiangsu Province (PRC), have introduced new digital devices to monitor the yarn quality of each bobbin, this concept has not been widely used. Hence, many industries neglect this problem and compromise the yarn quality significantly. The image of a ring spinning frame has been shown in Figure 3.6.

To resolve the above-mentioned problem, an RFID-based tracing system was designed by Liu et al. (Liu et al., 2010), which can assist to trace the bobbin to spindle of the ring frame during the yarn manufacturing process. RFID tags were successful to manage the bobbin to spindle data, which was

Figure 3.6 Ring spinning frame.

Source: Courtesy of Mr. Jyoti Ranjan Dash.

recorded in the ring spinning machine via dual-directional data communication. The information on a bobbin to spindle was managed via the database technology using computers. The collected information on the bobbin to spindle can be monitored to achieve production of yarn with better quality. The bobbin system with RFID tags is shown in Figure 3.7.

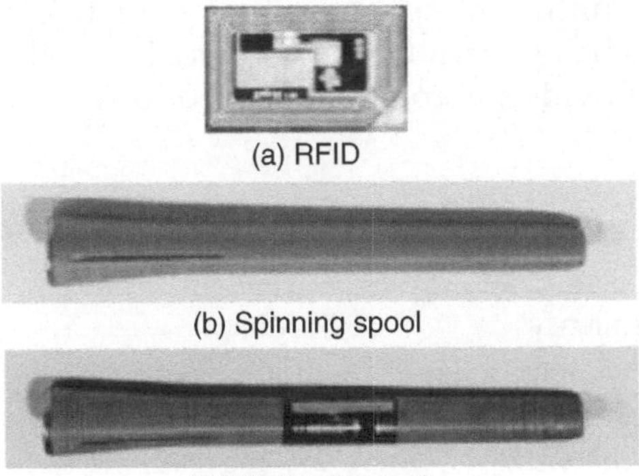

Figure 3.7 The ring frame bobbin (spool) with RFID tag: (a) RFID tag, (b) spinning spool or bobbin and (c) electronic spool. (Liu et al., 2010).

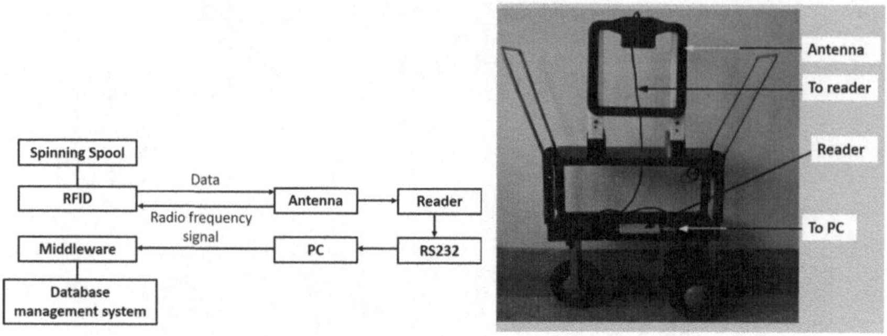

(a) (b)

Figure 3.8 Bobbin management system: (a) system structure and (b) hardware of the system (Liu et al., 2010).

The RFID tag was installed in the electronic spool which can help to identify the spindle with which the spool was mounted. The bobbin management system is an RFID-enabled database management system, which can be used to trace the information. In broader terms, the bobbin management system can effectively manage the tracing of bobbins with corresponding spindles. The system structure and the hardware of the system used for bobbin management are shown in Figure 3.8. Table 3.2 shows the specifications used for the bobbin management RFID system.

3.3.2 *Yarn quality tracking*

In the spinning industry, yarn is collected on bobbins mounted to spindles of a ring spinning frame. The spindles rotate at very high speed while the yarn is being manufactured. There is a sophisticated driving mechanism to provide the rotational speed to the spindles. At times,

Table 3.2 Specification of RFID bobbin management system (Liu et al., 2010)

Frequency	13.56 MHz
Read range	10–20 cm
Read rate	50 tags/sec
Memory size	64–256 bits read/ write
Power source	Inductive/ magnetic field
Advantage	Low-cost and standard frequency

the driving mechanism fails to work in the right manner, which leads to eccentric or inappropriate spindle rotation. This in turn leads to poor yarn quality due to non-uniform yarn. As the modern ring frame houses more than 1000 spindles, it is very hard to keep the track of each spindle and uniquely identify the faulty spindle or a group of faulty spindles.

As the testing of yarn does not include 100% inspection, it is hard to detect the faulty bobbins. Although the faulty bobbins can be detected in the subsequent process of winding, due to excessive end breakage or machine stops, it is hard to identify the spindle in the ring frame causing the fault. To solve this problem an electronic system is needed which can identify the bobbin and the corresponding spindle. Thus, it will be easier to locate the faulty spindle and fix the problem.

Shuai et al. (2010) used RFID technology to uniquely identify the faulty spindle that produced poor quality yarn during the spinning process in a ring spinning frame. They designed a low cost, easy implantable and high reliable system (shown in Figure 3.9), which works efficiently and effectively to detect the spindle faults. The RFID network was designed to store the ID of each bobbin in a computer database. Each bobbin was attached with an RFID chip and the reader was installed in the guide rail of the ring frame.

The ID of each bobbin was fed to the computer database by the reader. An EM4100 noncontact chip was used in combination with the U227013-STC89C52 reader integrated with a wireless module. In the case of any

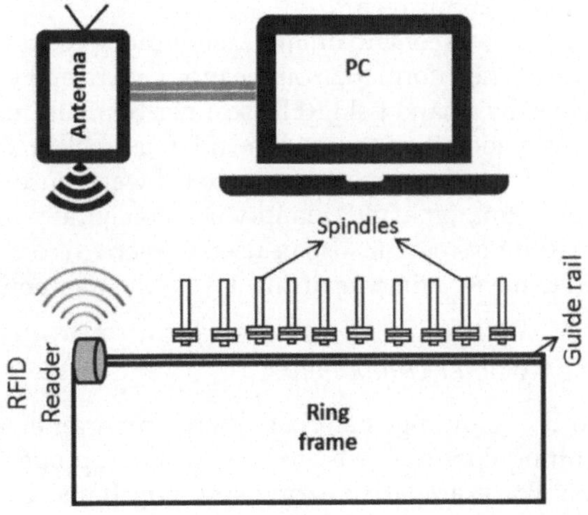

Figure 3.9 The design of RFID based system for yarn quality tracking.

spindle problem such as eccentric movement or spindle slippage, the bobbin can be identified easily and the yarn on the bobbin can be tested for quality consistency. This system effectively reduced the problem of poor quality yarn tracking and the bobbin identification problem in the ring spinning process.

3.3.3 Materials management

In yarn manufacturing industries, several fibers are used in the blow room, which finally converted into yarn. A specific type of fiber, for example cotton can vary in its fineness, color, tenacity and length (Nayak et al., 2015). It is difficult to differentiate the varieties of a fiber without an objective test. Hence, while manufacturing the yarn, it is essential the right fibers are used in the blow room. The recent practice of managing the fiber is manual, where the warehouse issues the fibers as per the stock keeping unit (SKU). However, there is a chance of the wrong fiber being selected due to human error. This can lead to different yarns than the specification.

Similarly, during the subsequent processes when slivers and roving are formed, there is a high chance of inter mixing of these materials due to similarity in the appearance of different materials. In a spinning industry, there are different types of spinning machines working on different fibers and yarn counts, such as rotor spinning, ring spinning and airjet spinning. Hence, there is a high chance of mixing different slivers and roving. The spinning frames manufacturing yarn are also vulnerable to mixing yarns of different counts. However, the necessity is to uniquely identify each yarn count in order to avoid mixing. The use of RFID technology can provide a great help in this regard.

As the RFID tags will contain unique identification codes for each material: fiber and yarn. The information on the type of yarn and yarn count can be easily identified by a hand-held RFID scanner. Similarly, the roving and slivers can be easily identified to avoid the mixing problem. Once the yarn is spun, it is sent to winding department or to the warehouse. The information on the yarn count, type and quantity is automatically updated using RFID tags. The RFID system can also be used to keep a record of yarns sold or yarns issued to the weaving department from the warehouse.

3.3.4 Managing other components

In addition to the spinning materials, there are several other components in a spinning floor that need to be selected in proper combination or specification. For example, in a ring spinning frame, a specific ring-traveler combination is needed for a specific yarn count. A change in the combination may lead to frequent end-breakages lowering productivity and efficiency. Similarly, for different counts of yarn, different gears are needed, which are managed manually in many industries. Manual management of these components is time consuming and error-prone. Hence, the use of RFID technology can help to select the right type of components within a small-time frame.

3.3.5 *Machine maintenance*

In a spinning industry there are several types of machines, including blow room, carding, draw frame, comber, speed frame, roving frame and ring frame. Each machine needs regular care and maintenance for proper working. There are hundreds of spare parts needed for the maintenance of the spinning machineries, which is difficult to manage with a manual approach. The use of RFID technology can help to keep accurate information about spare parts relating to any specific machine type and to manage their storage, issue and procurement.

3.4 *Conclusion*

The global spinning industries are heading towards modernization and automation. Fiber production facilities are also trying to implement new technologies for tracking and tracing. Many industries are also approaching to achieve the triple bottom line of sustainability (which are environmental, social and economic) in the manufacturing and supply chain processes by reducing material and process waste. The use of RFID technology can help in all these aspects with accuracy of information, automatic working and real-time information availability. The availability of real-time information saves the manual time-consuming process of receiving the information, which avoids many faults in the assembly line. In spinning industries, RFID technology is used for varied applications such as bobbin management, yarn quality tracking, material management and machine maintenance management.

In fiber production, RFID technology is being increasingly used. Various natural and synthetic fiber producers are realizing the benefits of RFID technology and adopting it. RFID technology is used in a wide range of applications during plant and animal fiber production such as: fiber bale tracking, synthetic fiber manufacturing and managing fiber testing. Governments in many countries are facilitating the use of RFID technology by providing the tags and the RFID system in subsidized prices, especially for wool fiber producers. In future, more and more fiber growers will be implementing this technology to achieve the benefits.

References

Age, C. S. 2005. Research Confirms Walmart's RFID Benefit. *Chain Store Age*, 81, 80.

Annabelle, C. 2017a. *Sheep RFID technology chaos* [Online]. Available: https://www.farmonline.com.au/story/4452488/sheep-rfid-technology-chaos/ [Accessed 18 July 2018].

Annabelle, C. 2017b. *WA to trial RFID tracking of wool bales* [Online]. Available: http://www.farmweekly.com.au/news/agriculture/sheep/wool/wa-to-trial-rfid-tracking-of-wool-bales/2755270.aspx [Accessed 15 June 2018].

AWTA. 2018. *Key Test Data Summary for June 2018* [Online]. Available: https://www.awtawooltesting.com.au/index.php/en/about/news/187-key-test-data-summary-for-june-2018 [Accessed 12 July 2018].

Brown, D., Savage, D., Hinch, G. & Hatcher, S. 2015. Monitoring liveweight in sheep is a valuable management strategy: a review of available technologies. *Animal Production Science*, 55, 427-436.

Claire, S. *Australian cotton gin tracks product to boost efficiency* [Online]. Available: http://www.rfidjournal.com/articles/view?13481/2 [Accessed 10 May 2018].

Christopher, M. 2016. *Logistics & Supply Chain Management*, UK, Pearson.

Dave, F. 2008. Australian sheep farmers explore RFID's benefits [Online]. Available https://www.rfidjournal.com/articles/view?4203 [Accesses 12 May 2018].

Friedlos, D. 2008. *Australian Sheep Farmers Explore RFID's Benefits* [Online]. Available: https://www.rfidjournal.com/articles/view?4203 [Accessed 10 July 2018].

Gereffi, G., & Frederick, S. 2010. The global apparel value chain, trade and the crisis: challenges and opportunities for developing countries. The World Bank.

Gupta, V. B. & Kothari, V. K. 2012. *Manufactured Fibre Technology*, Springer Science & Business Media.

John, H. 2008b. Ear tag [Online]. Available: https://en.wikipedia.org/wiki/Ear_tag [Accessed 20 July 2018].

Joshi, R. N. & Singh, S. P. 2010. Estimation of total factor productivity in the Indian garment industry. *Journal of Fashion Marketing and Management: An International Journal*, 14, 145-160.

Kadarusman, Y. & Nadvi, K. 2013. Competitiveness and technological upgrading in global value chains: evidence from the Indonesian electronics and garment sectors. *European Planning Studies*, 21, 1007-1028.

Lan, T. & Zhu, S. 2014. Chinese apparel value chains in Europe: low-end fast fashion, regionalization, and transnational entrepreneurship in Prato, Italy. Eurasian Geography and Economics, 55, 156-174.

Liu, J., Gao, W., Wang, H., Jiang, H. & Li, Z. 2010. Development of bobbin tracing system based on RFID technology. *The Journal of the Textile Institute*, 101, 925-930.

Maselyne, J., Saeys, W. & Van Nuffel, A. 2015. Quantifying animal feeding behaviour with a focus on pigs. *Physiology & Behavior*, 138, 37-51.

Nayak, R., Chatterjee, K., Khandual, A. & Gupta, T. 2007a. Scope & limitations of RFID. *Indian Textile Journal*, 117, 25-29.

Nayak, R., Chatterjee, K., Khurana, G. & Khandual, A. 2007b. RFID: Tagging the new era. *Man-Made Textiles in India*, 50, 174-177.

Nayak, R. & Padhye, R. 2014. Introduction: the apparel industry. *In*: Nayak, R. & Padhye, R. (eds.) *Garment Manufacturing Technology*. Elsevier.

Nayak, R. & Padhye, R. 2015. *Garment Manufacturing Technology*, Cambridge, UK, Woodhead Publishing.

Nayak, R. & Padhye, R. 2017. *Automation in Garment Manufacturing*, Woodhead Publishing.

Nayak, R. & Padhye, R. 2018. Introduction to automation in garment manufacturing. *In*: Nayak, R. & Padhye, R. (eds.) *Automation in Garment Manufacturing*. Elsevier.

Nayak, R., Singh, A., Padhye, R. & Wang, L. 2015. RFID in textile and clothing manufacturing: technology and challenges. *Fashion and Textiles*, 2, 1-16.

Nivaashini, M., Soundariya, R. & Kumar, D. 2018. Silkworm Growth Monitoring Smart Sericulture System based on Internet of Things (IOT) and Image Processing. *Network*, 180, 29-33.

NLIS. 2018. *National Livestock Identification System* [Online]. Available: https://www.animalhealthaustralia.com.au/what-we-do/biosecurity-services/national-livestock-identification-scheme/ [Accessed 1 July 2018].

Rain-RFID. 2017. *Sheep RFID Technology Chaos* [Online]. Available: https://rainrfid.org/member-news/2017/02/16/sheep-rfid-technology-chaos/ [Accessed 10 July 2018].

Ron, L. & Greg, S. *RFID Tag System for Cotton Gins* [Online]. Available: https://wefnexus.tamu.edu/rfid-tag-system-for-cotton-gins/ [Accessed 15 May 2018].

Saraswat, D. & Robbins, J. 2011. *RFID Technology May Improve Forecasting and Availability Reporting* [Online]. Available: https://www.nurserymag.com/article/nm1111-rfid-technology-tags/ [Accessed 7 August 2018].

Shaikh, T. N. & Bhattacharya, S. S. 2016. *Engineering Techniques of Ring Spinning*, WPI Publishing.

Shashikant, P. 2017. *RFID for animal identification in rural India and Asia Pacific* [Online]. Available: www.ieee-rfid.org/2017/03/28/rfid-for-animal-identification-in-rural-india-and-asia-pacific/ [Accessed 15 June 2018].

Shuai, W., Chongqi, M. & Hanming, L. Yarn quality tracking system based-on RFID. International Conference on Computer and Information Application (ICCIA), 2010 IEEE, 103-105.

Thomassey, S. 2010. Sales forecasts in clothing industry: the key success factor of the supply chain management. *International Journal of Production Economics*, 128, 470-483.

Van Der Velden, N. M., Patel, M. K. & Vogtländer, J. G. 2014. LCA benchmarking study on textiles made of cotton, polyester, nylon, acryl, or elastane. *The International Journal of Life Cycle Assessment*, 19, 331-356.

chapter four

RFID in fabric manufacturing and chemical processing

KEYWORDS: *fabric manufacturing, chemical processing, material management, labor management, batch identification*

4.1 Introduction

In the fabric manufacturing process, different types of fabrics are manufactured by weaving, knitting, nonwoven, felting, tufting and braiding technologies (Adanur, 2000). Generally, for apparel clothing, weaving and knitting are the most widely used processes. In weaving process, two sets of yarns, known as warp and weft, are intermeshed to form the fabric, whereas in the knitting process, horizontal and vertical loops are interlooped to form the fabrics (Sinclair, 2014). Each process has its own pros and cons and the fabric qualities are different. The common features of woven fabrics are low stretchability, lighter weight, high crease formation and better dimensional stability. Similarly, the features of knitted fabrics include high stretchability, heavier weight, better crease recovery and low dimensional stability.

In weaving industries, various steps are followed for preparing fabrics, which include: warping, sizing, weaving and quality inspection. In the warping process, the weaver's beam is prepared containing the number of ends (warps) in the fabrics (Gandhi, 2012). The sizing process applies a protective layer on the warp yarns to prevent thread damage during weaving. The weaving process is carried out by several looms such as shuttle looms, airjet, Sulzer and waterjet looms. The loom-state fabric or greige fabric is prepared, which is ready for the subsequent processes. In knitting process, the yarn packages from spinning are directly used to make the fabrics using warp and weft knitting machines (Spencer, 2001). The fabric manufactured by knitting is also subjected to different finishing processes to enhance the functionality.

The process sequence followed in weaving and knitting industries is shown in Figure 4.1. In the weaving and knitting industries, the raw materials include a wide range of yarns, which differ in their structure, fineness, strength, color and material type (natural and synthetic). As many of the weaving industries have different types of fabrics produced by a

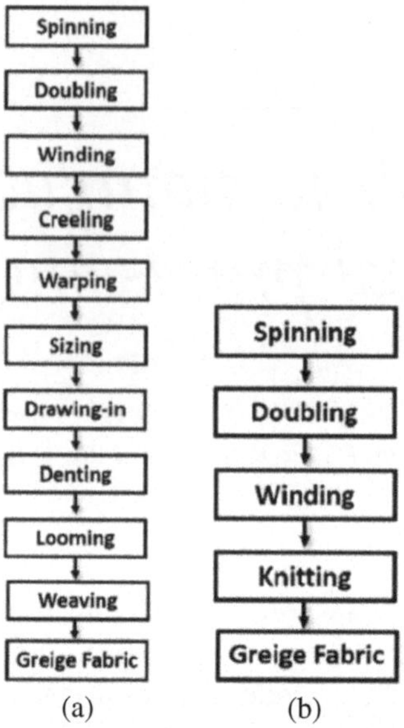

Figure 4.1 Process sequence in: (a) weaving and (b) knitting.

range of technologies, a wide range of materials are used for the machine maintenance. The recent trend in many weaving and knitting industries is still the manual practice of record keeping, which is time consuming and can be inaccurate. However, some industries are heading for automation to improve quality and productivity (Nayak and Padhye, 2017; Nayak and Padhye, 2018). The use of Radio Frequency Identification (RFID) technology can help in accurate record keeping and provide real-time information in the production floor (Nayak et al., 2007; Nayak et al., 2015).

The subsequent process in fabric formation is the chemical processing, where greige fabrics are processed by different chemical and mechanical processes to improve functionality and aesthetics (Schindler and Hauser, 2004). In chemical processing, various types of chemicals and auxiliaries are used for scouring, bleaching, mercerization, dyeing, printing, coating and other finishing applications. Many industries still use the traditional method of planning, resource allocation, material management, batch identification and information management (Oner et al., 2017).

This chapter discusses the application of RFID technology in various stages of weaving and knitting. This chapter also discusses various applications of RFID in chemical processing or wet processing, which is the most import step when value addition to fabrics is considered. Applications such as material management, labor management, machine maintenance, managing chemicals and auxiliaries, batch identification and routing process path are also discussed in this chapter.

4.2 RFID in knitting and weaving

As discussed, knitting and weaving processes are widely used for the manufacturing of fabrics for apparels. The use of RFID technology in various preparatory, weaving and warehousing processes has been discussed in this section.

4.2.1 Yarn management

The smart bobbin management system is shown in Figure 4.2. The RFID tags can be embedded into the cones or yarn packages used for knitting or weaving (Simonis et al., 2016). When the bobbins are loaded into the knitting or weaving machine, the data can be obtained by the RFID system. The details of the yarn parameters can be used for setting the machine parameters. This can reduce the machine setting time by 20% and improve the accuracy. This system can be used for large circular knitting machines, various types of looms and jacquard knitting technology. In addition to RFID system, the QR (quick response) codes can be used for storing the fabric information.

4.2.2 Material management

The conversion of yarn to fabric needs several processes and machines as discussed above. In each process there are a wide range of cones, other yarn packages, machine components and spare parts needed. The storage, identification and issuance of these materials is a time-consuming

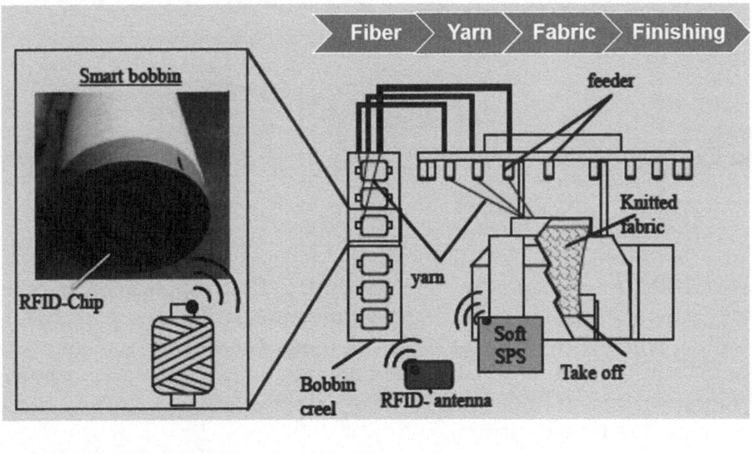

Figure **4.2** Smart bobbin system used for knitting.

Source: Simonis et al., 2016.

process. Although the use of computer-enabled software can help in keeping the records of the materials, managing the physical inventory and their movement is rather a difficult task. The store managers need to carefully identify and store or issue the correct parts. The use of RFID tags can make the process much simpler. In the manufacturing process of fabric, the inventory can be effectively managed using RFID tags.

In the warehouse management system (WMS) of weaving and knitting industries, the use of RFID tags can help to keep inventory records. For example, when the cones containing the RFID tags are taken from the shelves, stock is automatically reduced, and the record is updated. Similarly, in a fabric store when the fabric is sold and loaded in a truck to be delivered to the customer, the RFID readers record the data for the fabric being sold and automatically update the system. Hence, the RFID system can help to keep an accurate record and monitor the flow of material in the weaving department. These RFID applications adopted by Barburski et al. (2008) has been shown in Figure 4.3 and Figure 4.4. are shown in Figure 4.3.

In the traditional system of managing the inventory, the warehouse manager needs to physically enter the stock records issued to update the log, which is a time-consuming process. Furthermore, RFID

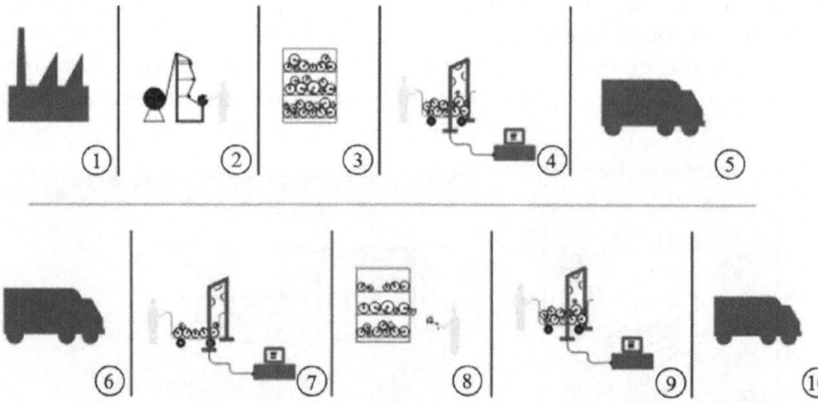

Figure 4.3 RFID application in textile industry. RFID application in a textile industry where: 1 – production, 2 – standardization section, preparation of fabric, labeling fabric with RFID tag, 3 – stock-room, 4 – RFID pike monitors fabric release, 5 – transportation of fabrics, 6 – delivery of fabrics to a wholesaler, 7 – acceptance of fabrics at RFID pike, 8 – wholesale stock-room, 9 – release of fabrics at RFID pike, 10 – transportation.

Source: Barburski et al., 2008.

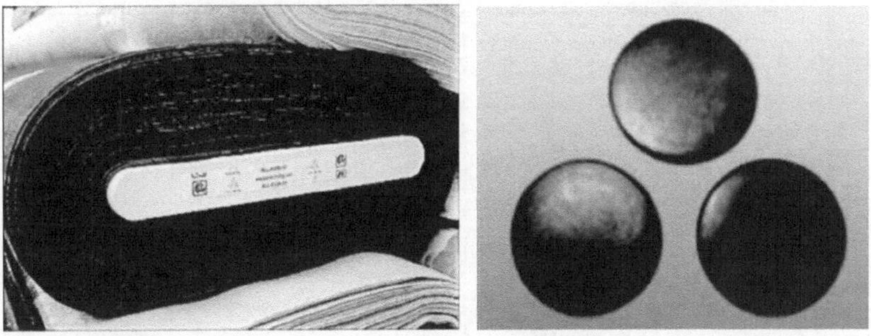

Figure 4.4 Fabric roll attached with RFID tags.

Source: Barburski et al., 2008.

systems can also prevent the theft of fabric or yarn in the warehouse, as unauthorized movement of the yarn packages or fabric rolls is recorded by the RFID system.

Various companies are also trying to implement an RFID system that will keep stock and assist in production control. For example, RFID tags have been applied by Griva (an Italian textile manufacturer) for its manufacturing plants. The use of RFID tags in fabric rolls to keep stock is shown in Figure 4.4 (Barburski et al., 2008).

Oner et al. (2017) investigated the benefits of implementing RFID technology in denim production plants. Denim production plants generally suffer from the problems of tracing, tracking and monitoring work in progress (WIP) inventories. On several occasions, incorrect record keeping, manual counting and barcode use lead to inaccurate physical stock and high labor cost. Furthermore, the traditional process is vulnerable to theft, incorrect product delivery and unexplained product loss.

The objective of their study was to collect real-time information on the manufacturing progress and malfunctions during the production process. Inaccuracy in record keeping and delivery were also taken into consideration using RFID tags. Furthermore, the authors have performed a detailed review of the literature during this investigation.

It was observed from the study that the use of RFID technology reduced the processing times for the denim manufacturing plant. In addition, the number of labor requirements was reduced by almost 50% while the RFID system was used. The RFID system was efficiently used to track malfunctions and problems that occurred on the production floor. This in turn helped to reduce product loss and penalty costs by 50%. The RFID network used is shown in Figure 4.5.

Although some textile industries are implementing the RFID technology on the production floor, it has not been widely adopted. In textile

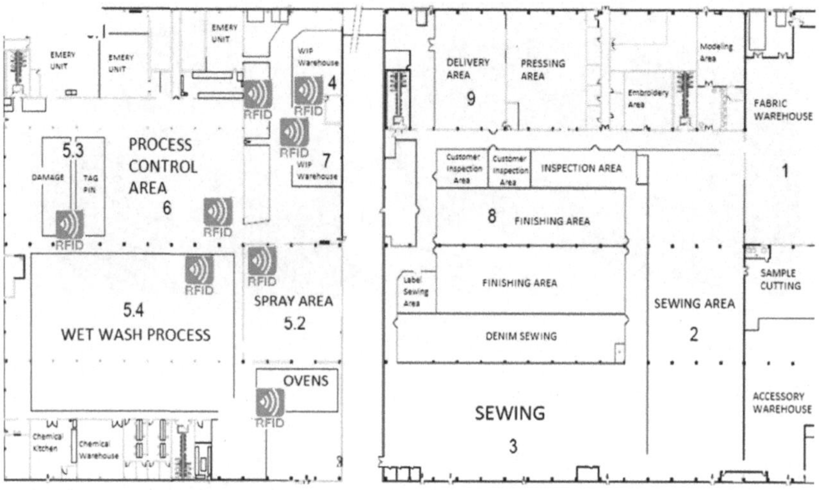

Figure 4.5 RFID Network used in garment industries.

Source: Oner et al., 2017.

manufacturing industry, it is hard to implement RFID technology due to the following reasons:

- Lack of standardized processes in many textile industries.
- Wet processing limits the use of RFID tags, which can be damaged by water and chemicals.
- Mechanical processes that can do physical damage to the tags.
- Several variations of final product specification (i.e., garment), which includes style, design, size and color.
- Similarly, there are a wide range of fiber, yarn and fabric types, which necessitates the use of a range of RFID tags depending on the area of use.
- The cost of the garments needs to be kept as low as possible. However, the use of multiple RFID tags can increase the product cost.
- RFID tag selection is one of the most challenging tasks, as the performance and cost depends on the tag.

4.2.3 *Machine maintenance*

In a weaving industry there are a wide range of machines for preparing the warp threads. The flow of production in the weaving floor has been shown in Figure 4.1, which includes warping or sectional warping, sizing, drawing-in, denting, looming and fabric production by various types of machines. Each machine needs regular care and maintenance for proper working. The use of RFID technology can help in keeping the maintenance schedule for these machines (Zhong et al., 2013). There are hundreds of spare parts needed for the maintenance of the weaving machines, which is difficult to manage with a manual approach. The use of RFID technology can help to keep accurate information about the spare parts relating to any specific machine type and to manage their storage, issue and procurement. Similarly, the fabric knitting machines and their corresponding maintenance and spare parts can be managed by RFID technology.

4.2.4 Production and productivity management

A fabric is manufactured by several operators on the weaving floor. With the use of RFID technology, the details of the operators, work done on the fabric by each operator, the loom number, loom type and date of manufacturing can be stored in an RFID tag. This set of information can be used when needed during the quality control and production planning and control (PPC) activities to track the operator or the machine.

In the stores of the weaving or knitting departments, stock levels can be quickly monitored with RFID technology, which helps in with real-time and efficient inventory management (Wang et al., 2006; Lee et al., 2013; Jihui et al., 2011). The location and movement of components as well as semi-finished and finished products can be detected by the RFID system, which helps in production monitoring and control. The performance of departments and individuals can be evaluated by the rate of production in each department. This helps in improving the productivity and quality of the fabrics manufactured.

4.2.5 Fabric quality inspection

Fabric inspection and quality management are essential aspects of producing good quality garment, which the garment manufactures should strictly follow (Nayak and Padhye, 2015; Nayak and Padhye, 2014b). Many industries inspect the fabrics on lighted tables with a metering device. This method is managed by experts in the field, and the fault details are kept on written records maintained in a log book. However, RFID technology can be used to keep track of defects in a fabric and remedial actions can be taken to remove them. The amount of rework done on a defective fabric can also be saved in an RFID tag, which can help to grade the fabric using different grading systems. The fabrics with more faults than the tolerance limit can be rejected.

4.2.6 Labor management

Fabric manufacturing is a labor-intensive process, similar to spinning. The cost of labor contributes 20–30% towards the final cost of the fabric.

As the cost of labor is rising in many developing countries, proper utilization of labor is essential to achieve higher productivity. RFID is an efficient tool to manage the labor force in a weaving industry (Dutta, 2007). Each employee carries an identity card with an RFID chip. The time the worker enters and leaves, total amount of break time, machine details, total working time and total absences in a month can be easily managed by an RFID tag. The information stored in an RFID tag can be used to calculate the payments and wages of workers (Nayak et al., 2015). In addition, the labor allocation to various machines on the start of a shift can also be managed by shift supervisors using the information from RFID tags. The RFID system can also help in estimating the lead time, identifying efficient and skilled employees and assisting in production planning and control (PPC) in the weaving department.

4.2.7 Fabric storage

Once the fabrics are manufactured, they are stored in a warehouse. The traditional approach is to allocate SKUs (stock keeping units) to each lot and store them in a designated area of the warehouse. Similarly, when the fabric is needed for a delivery, it is located manually from a log book, which is a time-consuming process. This process is managed manually by the warehouse manager and other employees. The use of RFID technology can make the process automatic and simpler (Choy et al., 2009). When the fabric enters the warehouse, the RFID tags can be scanned automatically by the RFID readers and the fabric details can be stored in the WMS. The use of RFID system can also help to uniquely identify the fabric type during the dispatching process and automatically update the stock into the WMS.

4.3 RFID in fabric chemical processing

Fabric chemical processing also known as wet processing involves the treatment of the greige fabric produced by weaving or knitting with different mechanical and chemical finishes enhancing the aesthetics and performance (Schindler, 2004). The fabric produced after weaving or knitting is not suitable for use, as it lacks aesthetic appeal, is vulnerable to high dimensional changes, has poor surface ornamentation, and is not easily converted to a three-dimensional garment. Hence, the fabric is subjected to various mechanical and chemical finishes to avoid these problems and make the garment making process easier.

Different chemical processing steps include singeing, scouring, bleaching, mercerization (for cotton fabric), dyeing and printing (Leube, 2002). The operation sequence for fabric chemical processing is shown

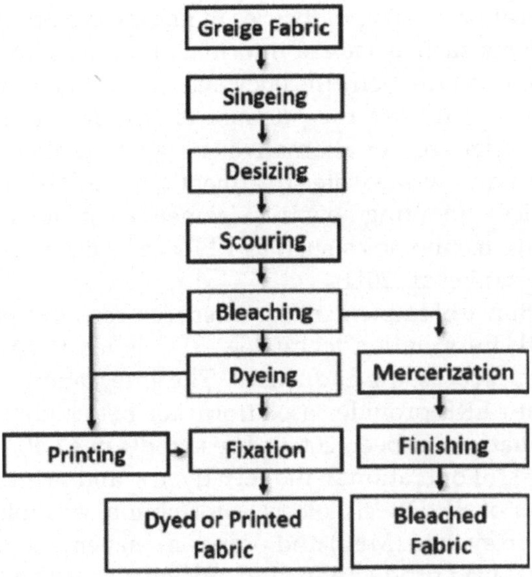

Figure 4.6 Sequence of operations in fabric chemical processing.

in Figure 4.6. Furthermore, various finishing processes such as wrinkle resistance, antibacterial, hydrophobic, flame retardant and soil release treatments are applied to the fabrics to enhance the performance to suit specific conditions (Nayak and Padhye, 2014a; Nayak et al., 2008). In all these processes different chemicals and auxiliaries are used mainly in a water-based medium that creates a large amount of waste water.

The textile chemical processing industry is the second most polluting sector in the world next to the oil sector. The major concerns are: air pollution due to the emission of carbon dioxide (CO_2) and other greenhouse gases (GHG); water pollution by the discharge of toxic effluent without treatment; and land pollution (Roy Choudhury, 2013). When environmental sustainability is considered, many of the chemical processing industries in developing countries fail to meet the global norms relating to environmental pollution. In addition, many industries take advantage of the lax rules and do not follow regulations relating to the environmental pollutions.

In many developing countries, the majority of the chemical processing industries work as separate units. They receive greige fabric from the weaving and knitting industries, perform the specific chemical processing operations and then return the processed cloth to the supplier. There are some composite industries with their own spinning, weaving and chemical processing facilities. These industries perform the chemical processing according to the fiber and fabric type and supply finished fabric to the garment industries in colors or prints as required.

The chemical processing industries in many developing counties suffer from problems such as lack of information availability from the production floor, low production efficiency and high energy consumption. In many developing countries the pressure of sustainable manufacturing, energy conservation, use of alternative energy, lowered emission levels, waste treatment and waste water treatment have led to changes in management practices, adopting new management concepts. Hence, the concept of adopting technologies such as RFID can help in achieving these objectives (Legnani et al., 2011).

The importance of implementing enterprise resource planning (ERP) integrated with information technology (IT) is felt necessary in many of the chemical processing industries. The integration of information technology with ERP provides a platform for better flow and management of information (Cebeci, 2009). These tools need RFID technology for their successful operation in modern dyeing and printing industries. The integration of RFID technology, information technology, manufacturing execution system (MES) and proper management of ERP systems can lead to increased productivity and efficiency; better enterprise management; and enhanced global competitiveness. The following section discusses the application of RFID technology in different sections of a textile chemical processing plant.

4.3.1 Managing chemicals and auxiliaries

In wet processing, hundreds of chemicals such as dyes, pigments, acids, bases, salts, bleaching agents and auxiliaries are used for different chemical processing steps. These chemicals are stored in chemical storage rooms in the textile industry. The recent practice is to allocate different SKUs manually to different chemicals and store them alphabetically according to their category, such as hazardous, flammables, toxic and corrosives. Many of the chemicals have certain periods of usable life after which they lose their strength or chemical reactivity. Hence, the manual method of managing the chemicals and using them before their expiration is a rather difficult process.

The use of RFID technology in chemical storage can help to eliminate the problem of managing hundreds of chemicals and their expiry dates (Barburski et al., 2008). When the chemicals are received, each one can be attached with an RFID tag containing the information. This helps to automatically update the stock report and allocate the shelf in the chemical store. Once they are stored, it is easier to identify them whenever they are needed for use. Furthermore, the storage alarm system can send a message before the expiry date of the chemical. Hence, many of the chemicals and auxiliaries can be used before their date of expiry, which can save a lot of money.

4.3.2 *Routing process path*

The current trend to establish a routing or process sequence for different types of fabrics is manually managed by the finishing managers and supervisors. They collect information from the logbook on the fabric type, inspect the fabric and finalize the process route. However, with the use of RFID technology, each fabric in the chemical process will contain an RFID tag. The processing managers and supervisors can know the type of fabric, quantity and history of the preparatory processes. Hence, they can use this data to decide the final routing for different fabrics. The traditional process of manual information collection, assigning machines and labor is eliminated and RFID tags make the process much simpler and more accurate.

4.3.3 *Assisting in pricing*

The use of RFID technology can provide real-time information on the usage of various machines, chemicals and auxiliaries for a specific type of fabric in dyeing and printing industries. The RFID technology can provide real-time information on the process sequence for each fabric and the amount of time taken in each process. Hence, the cost of dyeing and printing per linear meter of fabric can be calculated easily by the costing department by considering the energy, chemical, labor and material cost.

4.3.4 *Batch identification*

The dyed fabrics produced in different batches often vary in color values (CIE L*a*b (Commission Internationale de l'Elcairage)), although it is very minute (Luo et al., 2001). Mixing of different batches during the garment manufacturing process will lead to mismatches of the color in the different patterns of a garment. Hence, it is essential that the fabric manufactures keep each batch separate from the others and provide the information to the garment industries. The recent practice is to manually allocate SKUs to each batch and store them in the fabric warehouse. While delivering the fabrics to the garment manufacturers, it is highly likely that there is mixing of different batches. Similarly, in the warehouse of the garment industry it is highly likely that different batches are mixed when the fabrics are stored and issued to the cutting department.

The use of RFID technology can help to resolve these issues relating to the batch mixing. Each batch can be tagged with RFID tags containing unique codes that represent separate colors, batch numbers, dates, quantity of fabric and other material details. When they are stored at the fabric storage place, the information is readily updated into the store record. Similarly, when the fabric is issued, the RFID tags help in easy identification of all the fabrics of a batch and distinguish from the other batches. Hence, the RFID system can save time and avoid human error in mixing different fabric batches.

4.3.5 Increased visibility

In the chemical processing industry, each fabric lot needs to wait for some time due to the preparation of the machine for subsequent processing. The amount of down time before and after each process can be collected from the RFID tags used in each fabric lot. This data can inform the efficiency of each machine and identify bottlenecking in different production lines (Zhou, 2009). The production planning manager can monitor the work progress and verify that the timely completion of the work can be achieved or not. Furthermore, any faults generated during a specific operation can be identified and the reason can be easily investigated. This can help to increase production and customer satisfaction through timely delivery of finished fabrics to garment manufacturers downstream of the supply chain.

4.3.6 Increased labor productivity

Dyeing and printing industries are labor intensive. The workers in these departments are allocated jobs manually by their supervisors. At the beginning of a shift, the shift supervisor performs a head count and then allocates different machines to different operators. The use of RFID-based employee cards can help the supervisor to know the number of people available for each shift and allocate the machines accordingly. Similarly, a worker's temporarily absence from the work site can also be identified. Furthermore, depending on the process sequence, the amount of time taken by each machine can also be easily calculated, helping to calculate the efficiency of the workers as well as the machines. Hence, RFID technology can help to achieve increased labor productivity (Chow, 2006)

4.3.7 Labor wages and promotion

The use of RFID technology in the dyeing and printing industries can help to collect the number of hours a worker is working and their efficiency. This information can help the human resource department in managing payments, wages and leaves of employees. Furthermore, the RFID system can provide information on the efficiency of different workers, which can help to plan for incentives and promotions.

4.4 Conclusion

This chapter has focused on the use of RFID technology in the major fabric manufacturing processes such as weaving and knitting. In fabric manufacturing, RFID technology is used for inventory management, warehouse management, materials management, labor management, fabric inspection and storage. Due to increased labor prices and increased emphasis on fabric quality, the use of RFID technology can help to keep the cost of garments low. The benefits achieved by RFID technology can help to recover the high cost of implementation of the technology.

In fabric chemical processing, RFID technology is used for material management such as dyes, pigments, chemicals and auxiliaries; routing process paths for differ types of fabrics; accurate pricing of dyed and printed fabrics; identification of fabric produced in different batches; enhanced visibility of each process; and increased labor productivity. Many of the dyeing and printing industries have realized the impact and benefits of RFID technology. Some of them have already adopted RFID technology due to increased pressure from upstream garment manufacturers at the direction of leading fashion brands. However, as RFID technology is expensive and needs a skilled workforce, it will take time for complete implementation of the technology in many of the fabric manufacturing and chemical processing industries.

References

Adanur, S. 2000. *Handbook of Weaving*, CRC press.

Barburski, M., Czekalski, B. & Snycerski, M. 2008. RFID technology in the textile industry. *AUTEX Research Journal*, 8, 92-96.

Cebeci, U. 2009. Fuzzy AHP-based decision support system for selecting ERP systems in textile industry by using balanced scorecard. *Expert Systems with Applications*, 36, 8900-8909.

Chow, H. K., Choy, K. L., Lee, W. B. & Lau, K. C. 2006. Design of a RFID case-based resource management system for warehouse operations. *Expert Systems with Applications*, 30, 561-576.

Choy, K. L., Chow, K. H., Moon, K. L., Zeng, X., Lau, H. C., Chan, F. T. & Ho, G. T. 2009. A RFID-case-based sample management system for fashion product development. *Engineering Applications of Artificial Intelligence*, 22, 882-896.

Dutta, A., Lee, H. L. & Whang, S. 2007. RFID and operations management: technology, value, and incentives. *Production and Operations Management*, 16, 646-655.

Gandhi, K. 2012. *Woven Textiles: Principles, Technologies and Applications*, Elsevier.

Jihui, G., Fuhai, L., Bing, D., Peng, W. & Lehong, H. A Design of RFID tag antenna for clothing. Cross Strait Quad-Regional Radio Science and Wireless Technology Conference (CSQRWC), 2011. IEEE, 1075-1077.

Lee, C., Choy, K. L., Ho, G. T. & Law, K. 2013. A RFID-based resource allocation system for garment manufacturing. *Expert Systems with Applications*, 40, 784-799.

Legnani, E., Cavalieri, S., Pinto, R. & Dotti, S. 2011. The potential of RFID technology in the textile and clothing industry: opportunities, requirements and challenges. In: Ranasinghe D., Sheng Q., Zeadally S. (eds) *Unique Radio Innovation for the 21st Century*. Springer Berlin, Heidelberg.

Leube, H. 2002. Textile dyeing. In: Hunger, K (eds.) *Industrial Dyes: Chemistry, Properties, Applications*, 339-425.

Luo, M. R., Cui, G. & Rigg, B. 2001. The development of the CIE 2000 colour difference formula: CIEDE2000. 26, 340-350.
Nayak, R., Chatterjee, K., Khandual, A. & Jajpura, L. 2008. Evaluation of functional finishes-An overview. *Man-Made Textiles in India*, 51, 131-135.
Nayak, R., Chatterjee, K., Khurana, G. & Khandual, A. 2007. RFID: Tagging the new era. *Man-Made Textiles in India*, 50, 174-177.
Nayak, R. & Padhye, R. 2014a. Antimicrobial finishes for textiles. *In*: Paul, R. (eds.) *Functional Finishes for Textiles: Improving Comfort, Performance and Protection Elsevier.*
Nayak, R. & Padhye, R. 2014b. Introduction: the apparel industry. *In*: Nayak, R. & Padhye, R. (eds.) *Garment Manufacturing Technology*. Elsevier.
Nayak, R. & Padhye, R. 2015. Care labeling of clothing *In*: Nayak, R. & Padhye, R. (eds.) *Garment Manufacturing Technology*. Elsevier.
Nayak, R. & Padhye, R. 2017. *Automation in Garment Manufacturing*, Woodhead Publishing.
Nayak, R. & Padhye, R. 2018. Introduction to automation in garment manufacturing. *In*: Nayak, R. & Padhye, R. (eds.) *Automation in Garment Manufacturing*. Woodhead Publishing.
Nayak, R., Singh, A., Padhye, R. & Wang, L. 2015. RFID in textile and clothing manufacturing: technology and challenges. *Fashion and Textiles*, 2, 1-16.
Oner, M., Ustundag, A. & Budak, A. 2017. An RFID-based tracking system for denim production processes. *The International Journal of Advanced Manufacturing Technology*, 90, 591-604.
Roy Choudhury, A. K. 2013. Green chemistry and the textile industry. *Textile Progress*, 45, 3-143.
Schindler, W. D. & Hauser, P. J. 2004. *Chemical Finishing of Textiles*, Elsevier.
Simonis, K., Gloy, Y. & Gries, T. Industrie 4.0-Automation in weft knitting technology. 48th Conference of the International Federation of Knitting Technologists (IFKT) IOP Conference Series: Materials Science and Engineering, 2016. IOP Publishing, 141, 1-11.
Sinclair, R. 2014. *Textiles and Fashion*, Elsevier Science.
Spencer, D. J. 2001. *Knitting Technology: A Comprehensive Handbook and Practical Guide*, CRC press.
Wang, Z.-G., Tang, R.-Z., Sheng, W.-L. & Wang, G.-D. Research on RFID-based production logistics management techniques with application in garment industry. International Technology and Innovation Conference, ITIC 2006. IET, 383-387.
Zhong, R. Y., Dai, Q. Y., Qu, T., Hu, G. J. & Huang, G. Q. 2013. RFID-enabled real-time manufacturing execution system for mass-customization production. *Robotics and Computer-Integrated Manufacturing*, 29, 283-292.
Zhou, W. 2009. RFID and item-level information visibility. *European Journal of Operational Research*, 198, 252-258.

chapter five

Applications of RFID in fashion manufacturing

KEYWORDS: *fashion manufacturing, design development, inventory management, resource allocation, sewing productivity*

5.1 Introduction

The global need for fashion and textiles is fulfilled by manufacturing facilities in developing countries, as it is not economically viable to produce cheaper clothes in developed countries (Gereffi and Frederick, 2010). The last few decades have witnessed the shifting of clothing production to emerging countries such as Bangladesh, Vietnam, China, Indonesia, India and Cambodia, where wages are lower (Mani and Wheeler, 1998). This has helped to keep the price of fashion goods low because of cheap labor cost. Developed countries on the other hand are now focusing on the manufacturing of technical textiles used for medical, automotive, protection, filtration and electronic application due to their high profit margin.

The manufacturing of fashion, textiles and fashion accessories is still labor intensive despite the availability of many automation tools (Nayak and Padhye, 2017a; (Glock and Kunz, 2000). Although apparel manufacturing is labor intensive, often there is a high demand on product quality. Hence, to fulfill the high-quality requirements, it is necessary that the labor-intensive processes are converted into semi-automatic or automatic processes accomplished by using computerized tools and digital technologies (Glock & Kunz, 2000; Nayak & Padhye, 2018a). The use of Radio Frequency Identification (RFID) technology is one of the most widely used computerized tools in manufacturing industries including fashion (Nayak et al., 2015; Nayak et al., 2007a). Although there is a great opportunity for the application of RFID technology in fashion, it has not been widely adopted due to issues such as high cost of implementation, attitude of factory owners to save the budget and availability of cheap labor (Nayak et al., 2007b).

The recent trend in fashion is moving from the traditional fashion towards fast fashion. Today's consumers prefer to change their clothing more frequently by adopting fast fashion, which is available at cheaper prices more frequently. The other characteristics of fast fashion are short life-cycles, faster supply chain, just-in-time manufacturing and impulse buying. The manufacturing and distribution of fast fashion creates tremendous pressure on all the role players in the upstream and downstream of the fashion supply chain (Tang, 2006). It leads to several uncertainties at the macro level in the fashion manufacturing industry, which needs to change its business strategies, adopt faster practices, establish closer relationship with suppliers and react promptly to market changes.

In garment manufacturing industries, raw materials such as fabrics, trims and accessories are purchased form upstream suppliers and used to manufacture the garments. The garment manufacturing process involves fabric inspection, spreading, cutting, bundling, sewing, pressing, inspection and packing (Figure 5.1) (Wong et al., 2000; Nayak and Padhye, 2015a). As there are many variables in the designing of garments such as style, size and color, there are several components that need to be controlled separately to avoid errors and uniquely identify each component.

In many developing countries, the same manual, labor-intensive clothing production practices are still followed after many years, rather than the use of automatic equipment, despite the benefits. The major reason is the high cost of implementing new technologies such as RFID. This chapter discusses the applications of RFID technology in fashion manufacturing including fabric inspection, spreading, cutting, garment manufacturing, quality inspection and packaging.

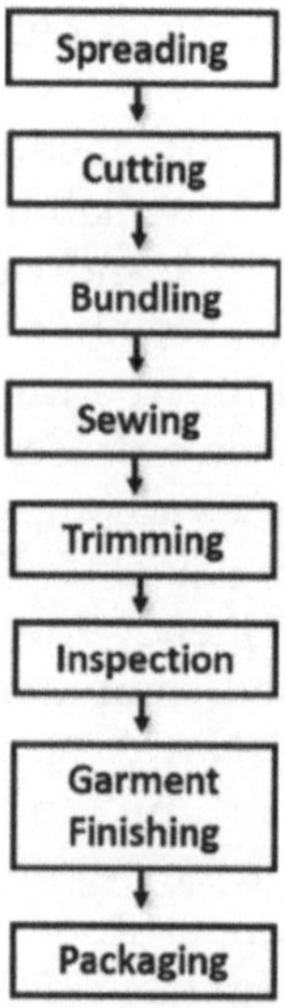

Figure 5.1 Process sequence of garment manufacturing.

Source: Nayak and Padhye, 2015b.

5.2 RFID in garment manufacturing

The garment manufacturing process involves several steps such as fabric spreading, cutting, bundling, sewing, pressing, inspecting and packaging as shown in Figure 5.1. It also involves design development, sample management and fabric inspection (Glock & Kunz, 2000). In each of these processes RFID technology can be uniquely applied to identify the components, accurately keep the stock record, track the objects and automatically update the information. The use of RFID technology increases the efficiency and the productivity of the garment manufacturing process, which are discussed in the following section.

5.2.1 RFID in design development

The design department is a vital part of any garment manufacturing industry. The designers and merchandisers prepare prototype samples of varying sizes for approval from the buyers (McKelvey & Munslow, 2011). They deal with several types of fabrics, trims and accessories on a daily basis. The use of RFID technology can help to keep track of different materials used for garment design development. Furthermore, additional information on the style, color and size can be stored in RFID tags. The number of style changes, the phase of a sample development and the work progress in the design room can also be monitored by RFID technology.

RFID technology can also help in asset management in a design department. For example, RFID systems can ensure the location of important assets in the design room such as dummies, sample garment pieces and location of laptops with specific software (CAD/CAM (computer-aided design/computer-aided manufacturing) or virtual prototyping). The dummies of different sizes are moved in the design section or sitting idle when not in use (Nayak and Padhye, 2017b). When needed, it is time consuming to locate a specific sized dummy. The RFID system can instantly locate the dummies. In addition, thousands of threads, buttons, zippers and labels are used in the design room, and it is rather a difficult task to locate. RFID technology can help in locating each component with accurate information. The cut fabric pieces in the design room can also be easily managed by an RFID system.

Each garment is constructed from basic pattern blocks. At the end of each order cycle, these blocks are stored in a storage place of the manufacturing industry. Over time thousands of these basic pattern blocks are accumulated in the storage place. On several occasions, there is a need to find the basic pattern blocks for the repeat orders, which is a difficult task. However, with the RFID tags, the process can be made simpler. Different bodice blocks can be grouped together and fitted with RFID tags, which makes it easier to find each block when needed.

5.2.2 Fabric inspection and storage

Fabric is the major component of a garment (about 50% of the cost of a garment) and it is the input material for many garment manufacturing industries. Once the fabric is received from the supplier, the quality of the fabric is inspected, and the length or weight is verified (Mak et al. 2009; Vijayan and Jadhav, 2015). RFID systems can help to uniquely identify the fabric type during the inspection and automatically update the stock into the warehouse management systems (WMS). Furthermore, any defects or faults in the fabric can be updated into the RFID tag. When the fabrics go to the cutting department, they can take care of the fabric faults by extracting the information from the RFID tags.

Once the fabric is inspected, it needs to be stored for a certain period before going to production. During storage, the RFID tags can identify the storage locations so that they are stored with similar type of fabrics. Furthermore, when they are taken from the storage to the conditioning room, they can again be easily identified. Once taken out of the storage place, the fabric record is automatically updated in the database.

5.2.3 Sample (fabric) management

The product development process in garment industries is quite time consuming and needs expert designers and merchandisers. This process includes forecasting, market research, trend analysis, sourcing and fabrication. The most important aspect is the selection of raw materials especially fabrics and trims. This process can be improved using support tools that can provide additional knowledge and improve the information access. The use of RFID systems in combination with artificial intelligence (AI) can help to select the right type of fabrics and trims.

Choy et al. (2009) investigated the scope of implementing an RFID-based sample management system for fabric selection during the design process. The RFID system investigated was interactive for the merchandisers and the designers to find the best way of selecting fabrics for designing new clothes and to expedite the product development process. The system provided appropriate knowledge to the designers and merchandisers for combining different fabrics effectively and efficiently. Furthermore, the RFID system provided real-time information on the fabric status in the storeroom and helped in tracking the fabric swatches. The working of the RFID system is explained in Figure 5.2.

The fabrics in the fabric-swatch storage cabinet were attached with passive RFID tag receivers. The passive RFID reader was placed at the top of the fabric-swatch storage cabinet, which was connected to a personal computer. The computer collected the information from the readers and

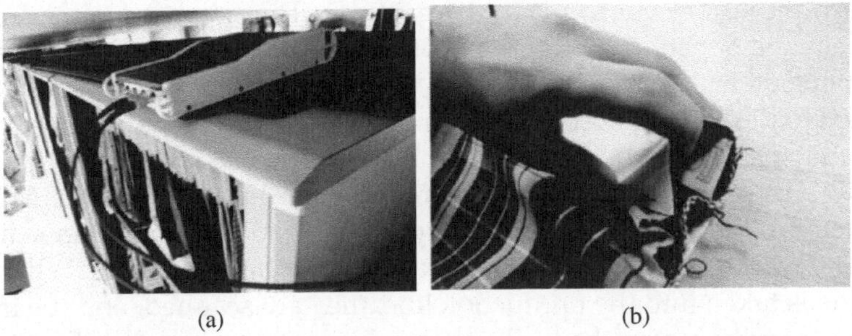

Figure 5.2 RFID system for fabric swatch selection for new product development: (a) RFID reader placed on top of swatch storage; and (b) fabric attached with the RFID tag.

Source: Choy et al., 2009.

reported to the designers on the status of each fabric. The hybrid case retrieval method was followed to retrieve the cases from the case repository to provide suggestions on fabric style, color and texture.

5.2.4 Fabric spreading and cutting

Fabric spreading and cutting are the most important steps influencing the quality of the final garment. Any mistakes during these steps are directly translated to the garment, which can lead to quality concerns or even rejections. The use of RFID technology can help in many ways to improve the accuracy and efficiency during spreading and cutting.

During spreading, different types of fabrics are spread on flat tables manually or automatically (Vilumsone-Nemes, 2015). It is important to take care of the fabric grain direction, type of spread (one-way, both-ways) and the fault present in the fabric. The information on the fabric quality, faults and lay directions can be extracted from the RFID tags while using a specific fabric lot. The important factor during spreading is avoiding lot mixing of fabrics. As there will be subtle color differences between different lots of fabric, the use of RFID tags can help to exactly identify the fabrics from the lot.

Without the RFID tags, the spreading team relies on pre-written instructions in log books or computer printouts, which is time consuming and error prone. The spreading team can get instantaneous information and take care of the quality of the spread with the use of RFID technology. Errors during spreading can be avoided using RFID tags, and the major problem of lot mixing can be eliminated.

The next step after spreading is cutting, where fabric patterns are cut by different cutting equipment. Depending on the garment style, several pattern pieces are cut to make the garment. These pattern pieces differ in their size depending on the size of the garment. After cutting, the cut components are bundled and fixed with a bundle ticket, which is then passed to the sewing floor. These bundle tickets are printed on paper that contains the information on the size, lot and garment style.

RFID tags containing the information can be used in the place of printed bundle tickets. Each RFID tag can be loaded with a unique number and attached to the fabrics similar to the printed bundles. When the bundle is taken into the production line, they are scanned, and the information is collected by the RFID system. The cut components of a specific garment can be loaded with a unique code to identify all the components. The advantages of using RFID tags include easy and accurate identification of the garment components. The other advantage is the tracking of the movement of each garment component in the sewing floor as discussed in the following section.

5.2.5 RFID in sewing floor

Sewing is a major step in garment manufacturing, which involves joining of different cut components of a garment by using a variety of sewing machines. Depending on the type of fabric (knit or woven), the sewing machines will be slightly varied. Skilled workers join different cut components in a specific sequence and perform other activities such as attaching the labels and trims (buttons, zippers and other closures) and trim the unnecessary threads.

There are three types of production systems in the garment industries: progressive bundle systems (PBS), unit production systems (UPS) and modular production systems (MPS) (Glock and Kunz, 2000). Among these, the PBS is widely used in many industries, which involves the progressive completion of the garment components from one operator to the other in sequence. The UPS involves the use of conveyor hanger systems, where the components are carried from one operator to another. This is an automatic practice for making the garments, which can save labor cost and time. Therefore, some industries are installing UPS. The MPS involves manufacturing the garment in groups and it is the least preferred method in the garment industries.

RFID technology can be applied to all three manufacturing systems described above. RFID technology can be effectively used to monitor material flow and information flow (Lee et al., 2013). RFID can improve complex operations on the sewing floor, which involves managing the movements of fabric components and trims. Each fabric component or trim will be fitted with an RFID tag, which is linked to a reader. The movement of each component from one operator to the other or from one machine to the other can be monitored, and the efficiency of the manufacturing system can be easily estimated. Any bottlenecking in the production floor can be easily identified when one of the components has stopped moving.

Hence, RFID technology is being used on sewing floors to obtain information on work progress, production quantities and to trace the efficiency of assembly lines. The other important aspect is the estimation of the efficiency of each sewing machine, productivity of each worker and overall productivity. In addition, RFID systems can detect which assembly line or specific worker on an assembly line was responsible for a defect. RFID systems can also track the status of each worker at any time and who is working in which style. Therefore, workers with minimal work can be allocated new tasks. Bottlenecks from the assembly line can be rapidly resolved due to rapid tracing of the causes and their sources.

Another important aspect of RFID technology is the accuracy of information flow (Zhou, 2009). The successful operation of an apparel industry relies on the effective flow of information from one department to another and how well it is handled and executed in real time. A reliable technological platform such as RFID is needed to fulfill these expectations. RFID technology can be used to collect and analyze the data during the manufacturing progression on a real-time basis. The common problems encountered in the production floor can be easily resolved using RFID technology.

5.2.6 *Marrying of wrong sizes*

In the garment industries, garments are manufactured in different sizes, using the progressive bundle system. As there are different sizes of garments, it is highly likely that different sizes of patterns will be mixed together creating a major problem of marrying of wrong sizes. This will cause fitting problems for the wearer. The use of RFID technology can help in avoiding this problem as the RFID tags can alert the sewing operator when wrong sizes are mixed during sewing.

5.2.7 *Increasing productivity*

RFID-based systems can be used to increase the productivity of garment manufacturing industries. For example, Lee et al. (2012) performed a case study in one of the largest garment manufacturers in Hong Kong using an RFID-based information system. They investigated the motivation behind the adoption of RFID technology in the manufacturing system and the success factors. This manufacturer knitted tops for kids, men and women for a number of famous fashion brands in Europe and the US. The industry uses barcode as well as RFID information systems to input operation performance daily basis. The manual system or the barcode system can't provide real-time data. Hence, with the manual system the production manager cannot keep track of the production progress, get information on the efficiency of production lines and number of defective products.

The manufacturing information system based on RFID technology consisted of: (1) a high frequency (13.56 MHz) passive tag with its unique ID associated with the bundle of cut-raw materials (such as sleeves, cuffs and hoods); (2) several RFID readers installed next to each sewing machine and quality control (QC) tables in the cutting department; and (3) a central PC workstation, connecting of 40 sets of RFID readers (the PC workstation collected the data first followed by uploading to the computer server); and (4) an Ethernet system to transmit data from PC station to the server.

In this case study RFID systems were installed in all sewing lines, and supporting staff were selected from Industrial Engineering Department (IED). The IED team was responsible for collecting data from different lines, analyzing and comparing the performance of each line and the performance without the RFID.

This study established the facts that: (a) the actual time for each operation in garment manufacturing can be easily recorded and this can be used for better line balancing; (b) this data can be used for establishing the standard time for each process and the workers can be allocated accordingly: and (c) the performance of individual workers can be monitored from the real-time data and the daily efficiency of each worker can be calculated. This data can be used to identify high-skilled laborers. This study also established the following tangible and intangible benefits of using RFID technology.

Tangible benefits

1. A 30% reduction in the *machine down-time* compared to last year.
2. A 50% reduction in the *lost-time* (non-working time) compared to last year.
3. A 30% reduction in the *defect rates* compared to last year.
4. A 50% increase in the *efficiency* compared to last year.
5. A 90% increase in *on-time delivery.*
6. A 10% reduction in the *amount of over-time* hours worked by sewers.

Intangible benefits

1. Uncover previously hidden problems, such as:
    - Co-ordination problem among departments.
    - Working efficiency of maintenance department.
    - Sewer's skill set deficiency and incompleteness.
2. Provide more accurate data for calculation of standard allowed minutes (SAM), which can be used for cost estimation and line-balancing.

3. Provide real-time production data for closer monitor of production defects, down times, efficiency, production line visibility and other issues such as process reengineering.
4. Build a culture of open discussion and scientific management, which is linked to the concept of continuous improvement.
5. Improve the work motivation of staff by creating an image that the company is willing to invest in helping the employees to enhance their work efficiency and obtain a higher payroll and bonus.

The demand for shorter lead time and more fashionable products with a variety of styles for each season is on the rise from the fashion retailers. To cater to the demand, the manufacturers always look for real-time monitoring systems that can react to deviations quickly and provide the necessary information. RFID technology can provide the solution to this challenge.

5.2.8 *Inventory management*

Inventory management is an important task in manufacturing industries including fashion and textiles (Moon and Ngai, 2008). While fashion manufacturing and retailing is being considered, inventory management is also rather complex due to short life cycles, several style variations and volatile demand. Profit maximization or stock turnover depends on the effectiveness of inventory management (Saygin, 2007). Insufficient inventory in manufacturing industry increases the chances of meeting the production schedule and the lead time. Similarly, excessive inventory will result in higher cost of inventory carrying cost.

In the recent trend of fast fashion, inventory management is more complex as the retail and manufacturing cycle together is five to six weeks (Jay et al., 2012). The allocation of resources (i.e., materials and machines) to a specific style in garment manufacturing industry is more challenging now. As the lead time is very short for fast fashion, allocating the number and types of machines, selecting a good marker and fixing the number of cutting pieces is managed by manual operations. The implementation of RFID technology is essential to make the process faster to cope with fast fashion manufacturing.

During the manufacturing of garments, the generation of waste is a common problem, especially fabric waste. As there is no standard method to predict the amount of waste for each production line, it is hard to determine the total amount of waste generated. Garment manufacturers purchase extra material (fabrics, linings and interlinings) anticipating 2-5% loss during the manufacturing process. However, it can be higher in some cases, which necessitates the purchase of fresh materials. Buying a fresh batch of a fabric lot is challenging as the color will vary from batch-to-batch. Hence, the problem of color variations will arise. The use of RFID technology can help in keeping accurate information on the batch or lot number and help in waste management.

An RFID system assists in improved inventory management and enhances customer shopping experience (Hardgrave et al., 2009; De Kok et al., 2008). RFID use also helps in reducing labor. For example, American Apparel applied RFID systems in eight of their stores, which saved about 60–80 hours per week in labor and reduced out-of-stock situations that arise due to insufficient information on volume of stock. Similarly, the Japanese apparel manufacturer Sankei is using an RFID system to track clothes during the manufacturing process and in online sales (Wu et al., 2009).

A successful case study has been conducted by Kaufhaulf departmental store in Europe on the impact of RFID on their processes and supply chain (Loebbecke and Huyskens, 2008). They have concluded that RFID can be used in warehousing to track goods and assist in cross-docking operations, as the goods received are not segregated in the warehouse but stored as they are received from the manufacturer. In the distribution center's receivable area, RFID readers can be implemented to track goods as they come in. The received goods are registered and the data are stored on RFID transponders allowing them to be read whether received in pallets or as individual fashion items on hangers. However, the task of counting hangers, items and pallets can be eliminated, and quantity can be controlled up to a 100% satisfaction level.

5.2.9 *Availability of real-time information*

Garment industries lack real-time information systems on the manufacturing floor. Most work-related information is managed by manual methods and log books. For example, the productivity of each production line is recorded every hour manually, which does not allow the production manager to identify the bottlenecks at an early stage. Hence, the lack of available timely information makes it difficult to monitor process delays or the quality of the product. Several garment industries use a paper ticket system to manage various operations and process delays. Hence, process delays are invisible to management until the garments are finished. The use of RFID technology on the production floor can help to deliver real-time information (Zhou, 2009) on bottlenecks and the work progress of each line, which can improve efficiency and productivity. RFID technology can be used to monitor operations on the manufacturing floor and avoid delays in meeting delivery dates.

5.2.10 *Resource allocation*

Resource allocation is always an important aspect of many manufacturing industries including garments. A proper resource allocation system can not only allocate resources but also increase profitability. In the garment industry, various resources include materials, machines and labor. Proper allocation of the resources is essential for increased productivity and efficiency by avoiding possible bottlenecks (Lee et al., 2013). This also provides competitive advantages to an industry. As the fashion industry is characterized by short cycles and volatile demand, effective resource allocation is an important aspect of responding to quick market changes.

In the garment industry, resource allocation is a rather difficult task as it involves hundreds of different cut components and trims; various types of machines; and a labor force with different skill sets. Even in today's garment industries this process is performed by human experts, which is time consuming and inaccurate. Furthermore, any delay in decision making or any error can cause delay in the production schedule. Several researchers have adopted approaches such as solving scheduling and line balancing (Wong et al., 2000); mathematical modeling (Guo et al., 2006); artificial intelligence and data mining (Elango et al., 2011; Ho et al., 2012) techniques.

Lee et al. (2013) designed an RFID-based Resource Allocation System (RFID-RAS) for real-time data capturing and resource allocation for garment manufacturing as shown in Figure 5.3. An artificial intelligence (AI)–based fuzzy logic network was used for the RFID-RAS system to provide human intelligence for real-time information (Nayak and Padhye, 2018b).

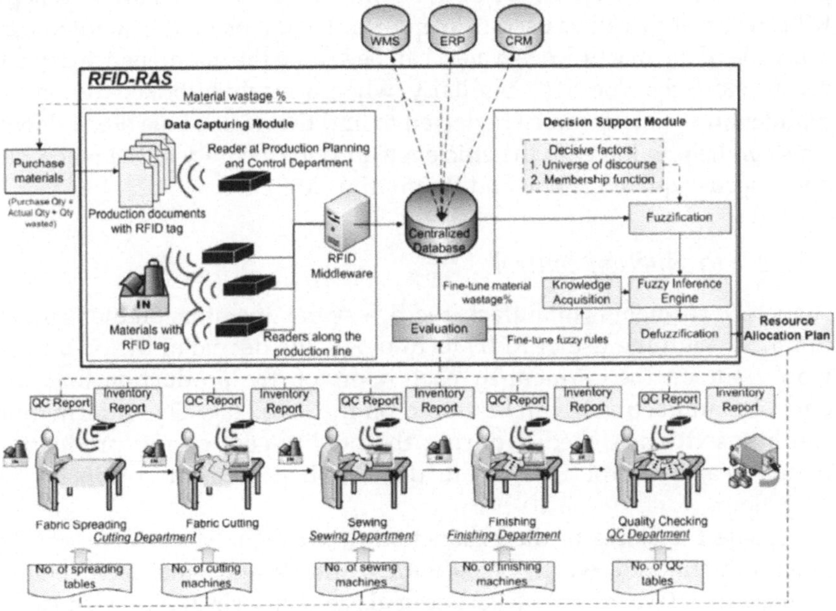

Figure 5.3 RFID-based resource allocation system.

Source: Lee et al., 2013.

The RFID-RAS system was used to regularly track real-time information regarding the actual material wastage. It was found that the RFID-RAS was very efficient and effective in providing real-time information for resource allocation.

The RFID-based resource allocation system consisted of two modules: (a) a data capturing module (DCM) and (b) a decision support module (DSM). The data capturing module helped in collecting real-time data relating to production, whereas the decision support module helped in resource allocation. The data were stored in a centralized database and linked to the WMS, enterprise resource planning (ERP) and customer relationship management (CRM). When needed, data can be collected from the DCM and transferred to the DSM. Fuzzy logic principles were used for resource allocation.

5.2.11 Maintenance of machines

The garment manufacturing process involves different types of machines for sewing such as lock stitch, chain stitch, button holing machines, button attaching machines; fusing and pressing equipment. All the equipment is maintained on regular basis so as to avoid major breakdowns on the sewing floor. The maintenance of the machines is done manually, and physical copies of the records of the maintenance schedule are also kept. RFID technology can be used to keep the information on the maintenance schedule of the machines. An alert can be sent to the maintenance department before the due date. Similarly, when a machine breaks down, its maintenance history can be reviewed to find the cause of the break down. This can help to increase the efficiency and productivity of the machines and improve quality (Zhou and Piramuthu, 2013).

5.2.12 Production control

A specific garment is manufactured by several operators on the production line. With the use of RFID technology, the details of the operators, work done on the garment by each operator, the production line and date of manufacturing can be stored on the RFID tag. This information can be used when needed during the quality control and production planning and control (PPC). It can also help in production control on the shop floor.

Inside the stores of the production department, stock levels can be quickly monitored, which helps with real-time and efficient inventory management (Wang et al., 2006; Lee et al., 2013, Jihui et al., 2011). The location and movement of the components and the semi-finished and finished products can be detected by the RFID system, which helps in production monitoring and control. The performance of departments and individuals can be evaluated by the rate of movement of the garment components between departments and individuals. This helps in improving the productivity and quality. On the packing floor the mixing of different styles and sizes can be avoided with RFID tags. In addition, the number of garment pieces in various packed cartons can be counted without opening, which can save time and labor.

5.2.13 *Quality management*

Garment inspection and quality management is an essential aspect of producing good quality garments. Many industries follow the concept of final inspection, where a selective number of garment pieces based on the acceptance quality level (AQL) are selected and inspected for faults (Keist, 2015). Some industries also perform in-process inspection after each manufacturing stage. The traditional practice of keeping the data is manually maintaining a log book. However, the use of RFID technology can keep track of defects in a garment and the remedial actions taken to remove them. The amount of rework done on a defective garment can also be saved using an RFID tag, which can help to pass or a fail a garment in the inspection. Garments with more faults than the tolerance limit are sold as seconds or rejected.

5.2.14 *Office management*

RFID technology can be used in the office management and central buying offices of the fashion brands. Various functions, such as movement of files, managing confidential documents and asset management in the office can be controlled by the RFID system. In the buying offices, the location of various design files can be traced by the RFID system, which can be used to monitor work progress. Similarly, sensitive files can be detected by an alarm system if they move out of the specified territory. Important assets such as printers, plotters and dummies in an office can be easily detected by the RFID system when they are idle, which can help in the efficient utilization.

5.2.15 *Labor management*

In garment manufacturing, the cost of labor contributes 30–40% towards the final cost of the garment (Singh and Nijhar, 2015). The cost of labor is rising in many developing countries due to minimum wage policy and stricter regulations by labor organizations. RFID is an efficient tool to help keep labor costs low (Jones et al., 2005). Each employee carries an identity card with an RFID chip. The time the worker enters and leaves the factory, total amount of break time, work-related injuries, total working time and total absences in a month can be easily managed with the RFID tag. This information can be used for calculating the payments and wages of the workers. In addition, the labor allocation in the beginning of a shift can also be managed by shift supervisors. The RFID system can also help in estimating lead time; finding efficient and skilled employees; total labor cost of a garment style; and assist in PPC.

5.2.16 Warehouse management

In garment manufacturing industries there can be two types of warehouses. The first type of warehouse is used to store raw materials such as fabrics and trims; maintenance, repair and operations (MRO) supplies. The second type of warehouse is used to store the finished garments before they are dispatched to the port of transportation or to their final destination. Managing the warehouse used for storing the finished garments is a rather complex activity, as there are as many varieties of garments as destinations. Without the RFID system, the garments are counted manually, allocated with some store keeping unit (SKU) and stored at the warehouse. Each time the garments leave the warehouse, the stock level is manually updated by the store manager or a warehouse employee. The storage of finished garments in warehouses is shown in Figure 5.4.

However, the use of RFID technology can make this process much simpler (Chen et al., 2013). When the final garments enter the warehouse, the information stored in the RFID tags is automatically scanned by an RFID reader and updated in the WMS. When the garments are taken for the final shipment from the warehouse, the information is also updated automatically. In addition, locating the garments during the storage is much easier with the RFID technology. The use of the RFID system increases the accuracy, saves time on locating the garments, and facilitates on-time delivery of the order.

5.3 Conclusion

RFID technology is becoming a popular tool for factory management in many global apparel industries. Several industries have already implemented RFID technology and getting the benefits of it. Other industries are at the stage of implementing the technology due to the mandatory requirements of many fashion brands. For small and medium enterprises, the installation of RFID technology is a real challenge due to the high cost of RFID tags and the RFID system. However, in the future, the help from government can boost the use of RFID technology.

During garment manufacturing there are various processes that use RFID technology including: fabric inspection and storage; fabric

(a)

(b)

Figure 5.4 Storage of finished garments in the warehouse of a garment industry: (a) garments in hanger storage and (b) garments in box storage.

Source: (a) & (b) Courtesy of Tho Nguyen Huu.

spreading and cutting; sewing operations; quality control, packaging and warehousing. The activities that are controlled by RFID technology include inventory management; real-time information availability; resource allocation; labor management; machine maintenance; production control; and tracking of objects. In the future, the cost of an RFID system will be affordable by small and medium-sized enterprises (SMEs), which will lead to its widespread acceptance. The industries that newly implement RFID systems should understand the actual benefits and success factors of an RFID system.

References

Chen, J. C., Cheng, C. H., Huang, P. B., Wang, K. J., Huang, C. J. & Ting, T. C. 2013. Warehouse management with lean and RFID application: a case study. *International Journal of Advanced Manufacturing Technology*, 69, 531-542.

Choy, K. L., Chow, K. H., Moon, K. L., Zeng, X., Lau, H. C., Chan, F. T. & Ho, G. T. 2009. A RFID-case-based sample management system for fashion product development. *Engineering Applications of Artificial Intelligence*, 22, 882-896.

De Kok, A., Van Donselaar, K. & Van Weonsel, T. 2008. A break-even analysis of RFID technology for inventory sensitive to shrinkage. *International Journal of Production Economics*, 112, 521-531.

Eelango, M., Nachiappan, S. & Tiwari, M. K. 2011. Balancing task allocation in multi-robot systems using K-means clustering and auction based mechanisms. *Expert Systems with Applications*, 38, 6486-6491.

Gereffi, G. & Frederick, S. 2010. The global apparel value chain, trade and the crisis: challenges and opportunities for developing countries, The World Bank.

Glock, R. E. & Kunz, G. I. 2000. *Apparel manufacturing: Sewn product analysis*, Prentice Hall Upper Saddle River, NJ.

Guo, Z., Wong, W. K., Leung, S., Fan, J. & Chan, S. 2006. Mathematical model and genetic optimization for the job shop scheduling problem in a mixed-and multi-product assembly environment: A case study based on the apparel industry. *Computers & Industrial Engineering*, 50, 202-219.

Hardgrave, B. C., Aloysius, J. & Goyal, S. 2009. Does RFID improve inventory accuracy? A preliminary analysis. *International Journal of RF Technologies: Research and Applications*, 1, 44-56.

Ho, G. T., Ip, W., Lee, C. & Mou, W. 2012. Customer grouping for better resources allocation using GA based clustering technique. *Expert Systems with Applications*, 39, 1979-1987.

Jihui, G., Fuhai, L., Bing, D., Peng, W. & Lehong, H. A Design of RFID tag antenna for clothing. Cross Strait Quad-Regional Radio Science and Wireless Technology Conference (CSQRWC), 2011. IEEE, 1075-1077.

Jones, P., Clarke-Hill, C., Hillier, D. & Comfort, D. 2005. The benefits, challenges and impacts of radio frequency identification technology (RFID) for retailers in the UK. *Marketing Intelligence and Planning*, 23, 395-402.

Joy, A., Sherry Jr, J. F., Venkatesh, A., Wang, J. & Chan, R. 2012. Fast fashion, sustainability and the ethical appeal of luxury brands. *Fashion Theory*, 16, 273-295.

Keist, C. 2015. Quality control and quality assurance in the apparel industry *In*: Nayak, R. & Padhye, R. (eds.). *Garment Manufacturing Technology*. Elsevier.

Lee, C., Choy, K., Law, K. & Ho, G. An intelligent system for production resources planning in Hong Kong garment industry. In Industrial Engineering and Engineering Management (IEEM), 2012. IEEE, 889-893.

Lee, C., Choy, K. L., Ho, G. T. & Law, K. 2013. A RFID-based resource allocation system for garment manufacturing. *Expert Systems with Applications*, 40, 784-799.

Loebbecke, C. & Huyskens, C. 2008. A competitive perspective on standard-making: Kaufhof's rfid project in fashion retailing. *Electronic Markets*, 18, 30-38.

Mak, K. L., Peng, P. & Yiu, K. F. C. 2009. Fabric defect detection using morphological filters. *Image and Vision Computing*, 27, 1585-1592.

Mani, M. & Wheeler, D. 1998. In search of pollution havens? Dirty industry in the world economy, 1960 to 1995. *The Journal of Environment & Development*, 7, 215-247.

McKelvey, K. & Munslow, J. 2011. *Fashion Design: Process, Innovation and Practice*, John Wiley & Sons.

Moon, K. L. & Ngai, E. W. T. 2008. The adoption of RFID in fashion retailing: a business value-added framework. *Industrial Management and Data Systems*, 108, 596-612.

Nayak, R., Chatterjee, K., Khandual, A. & Gupta, T. 2007a. Scope & limitations of RFID. *Indian Textile Journal*, 25-29.

Nayak, R., Chatterjee, K., Khurana, G. & Khandual, A. 2007b. RFID: Tagging the new era. *Man-Made Textiles in India*, 50, 174-177.

Nayak, R. & Padhye, R. 2015a. Introduction: the apparel industry. *In*: Nayak, R. & Padhye, R. (eds.) *Garment Manufacturing Technology*. Elsevier.

Nayak, R. & Padhye, R. 2015b. *Garment Manufacturing Technology*, Elsevier.

Nayak, R. & Padhye, R. 2017a. *Automation in Garment Manufacturing*, Woodhead Publishing.

Nayak, R. & Padhye, R. 2017b. *Manikins for Textile Evaluation*, Woodhead Publishing.

Nayak, R. & Padhye, R. 2018a. Artificial intelligence and its application in the apparel industry. *In*: Nayak, R. & Padhye, R. (eds.) *Automation in Garment Manufacturing*. Woodhead Publishing.

Nayak, R. & Padhye, R. 2018b. Introduction to automation in garment manufacturing. *In*: Nayak, R. & Padhye, R. (eds.) *Automation in Garment Manufacturing*. Elsevier.

Nayak, R., Singh, A., Padhye, R. & Wang, L. 2015. RFID in textile and clothing manufacturing: technology and challenges. *Fashion and Textiles*, 2, 1-16.

Saygin, C. 2007. Adaptive inventory management using RFID data. *The International Journal of Advanced Manufacturing Technology*, 32, 1045-1051.

Singh, A. & Nijhar, K. 2015. Garment costing. *In*: Nayak, R. & Padhye, R. (eds.) *Garment Manufacturing Technology*. Elsevier.

Tang, C. S. 2006. Perspectives in supply chain risk management. *International Journal of Production Economics*, 103, 451-488.

Vijayan, A. & Jadhav, A. 2015. Fabric sourcing and selection. *In*: Nayak, R. & Padhye, R. (eds.) *Garment Manufacturing Technology*. Elsevier.

Vilumsone-Nemes, I. 2015. Fabric spreading and cutting. *In*: Nayak, R. & Padhye, R. (eds.) *Garment Manufacturing Technology*. Elsevier.

Wang, Z.-G., Tang, R.-Z., Sheng, W.-L. & Wang, G.-D. Research on RFID-based production logistics management techniques with application in garment industry. International Technology and Innovation Conference, 2006. (ITIC 2006). IET, 383-387.

Wong, W., Chan, C. & Ip, W. 2000. Optimization of spreading and cutting sequencing model in garment manufacturing. *Computers in Industry*, 43, 1-10.

Wu, D.-L., Ng, W. W., Yeung, D. S. & Ding, H.-L. A brief survey on current RFID applications. International Conference on Machine Learning and Cybernetics, 2009. IEEE, 2330-2335.

Zhou, W. & Piramuthu, S. 2013. Remanufacturing with RFID item-level information: optimization, waste reduction and quality improvement. *International Journal of Production Economics*, 145, 647-657.

Zhou, W. 2009. RFID and item-level information visibility. *European Journal of Operational Research*, 198, 252-258.

chapter six

Application of RFID in fashion logistics

KEYWORDS: *fashion logistics, shrinkage, inventory management, warehouse management, managing returns*

6.1 Introduction

The process of strategically managing the efficient flow and storage of raw materials, in-process inventory and finished goods from point of origin to point of consumption is logistics activity. The fashion supply chain consists of network of organizations that are involved, through upstream and downstream linkages, in the different processes and activities that produce value in the form of products and services ultimately in the hands of the consumer (Christopher et al., 2004). The upstream players in the fashion supply chain are the fiber producers, yarn and fabric manufactures and finally the garment manufacturers. Previous chapters have discussed the use of Radio Frequency Identification (RFID) technology in the various upstream players up to garment manufacturing. This chapter will focus on the distribution and logistics activities once the garments are dispatched from the point of origin to the point of destination.

Fashion supply chain activities have undergone several changes due to the global shift of manufacturing from developed countries to emerging economies, which are thousands of kilometers away (Nayak and Padhye, 2015). Integrating all the upstream and downstream players in the supply chain is rather a difficult task due to the complex activities involved. To date, many of the fashion brands manage their supply chain activities manually. Inaccuracy and delay in information availability make the process rather difficult to control. The use of RFID technology in the fashion supply chain can provide up-to-date and accurate information, which is the key to efficiently managing the whole supply chain (Christopher and Peck, 1997).

Fashion supply chain is characterized by high levels of product differentiation, short life cycle, volatile consumer demand, low product cost, high level of impact buying, low predictability and high demand for quality (Nayak and Padhye, 2014; Masson et al., 2007). To be successful in the global competitive market, fashion manufacturers and retailers should focus on innovative products and processes in the whole supply chain. Many of the fashion brands now focus on: (1) adopting sustainable raw materials (which are derived from natural resources and are biodegradable); (2) recycling of plastics into textiles; (3) upcycling; and (4) adoption of automation (Nayak and Padhye, 2018a) and the application of technology such as the RFID.

Among these, the adoption of technology can help in several ways to keep production costs low. Due to stiff global competition, many fashion manufacturers have found it difficult to keep product prices low. Furthermore, the rise in material cost and increasing labor prices in many upstream and downstream activities in developing countries have accentuated the problem. Hence, the adoption of technology can help to keep the cost low. Moving towards automation can help to increase productivity, which in turn can assist to reduce the cost of production (Nayak and Padhye, 2017; Nayak and Padhye, 2018b).

The use of RFID technology in fashion or any other supply chain can help in several ways, such as easier materials management, effective inventory control, stock replenishment, prevention of theft, tracking of products and effectively managing the merchandising in retail stores. As mentioned earlier, RFID is a noncontact technology; it can detect product location and generate data without human intervention. This feature has enabled RFID technology to be widely adopted in the supply chain of many products including fashion.

Fashion garments are manufactured in various small, medium and large enterprises in developing countries. Form the developing countries (origin) they are transported to the port of developed countries (destination) through container ships. The garments are first transported by the truck to the port, unloaded from the trucks and kept alongside the ship. Subsequently, the garments are loaded into the containers; containers are then loaded into the ship and are transported to the destination. After the arrival of the ship at the destination port, the garments are transported to the distribution center (DC) of the fashion brand. The distribution center works as a warehouse for many retail stores. As per the requirements of individual retail stores, the number of garment pieces is delivered from the distribution center.

In many countries, the logistics process is managed by manual methods of record keeping. The export terms and conditions, such as Ex-Works (EXW), free-on-board (FOB) or cost, insurance and freight (CIF), are also kept manually, which is time consuming. The use of RFID technology can help to reduce the time spent and increase accuracy. The fashion logistics process is shown in Figure 6.1.

RFID systems were commercially introduced in the 1960s. However, the application of RFID systems in logistics management is quite new; it can be dated back to 2003, when it was mandated by global retailers such

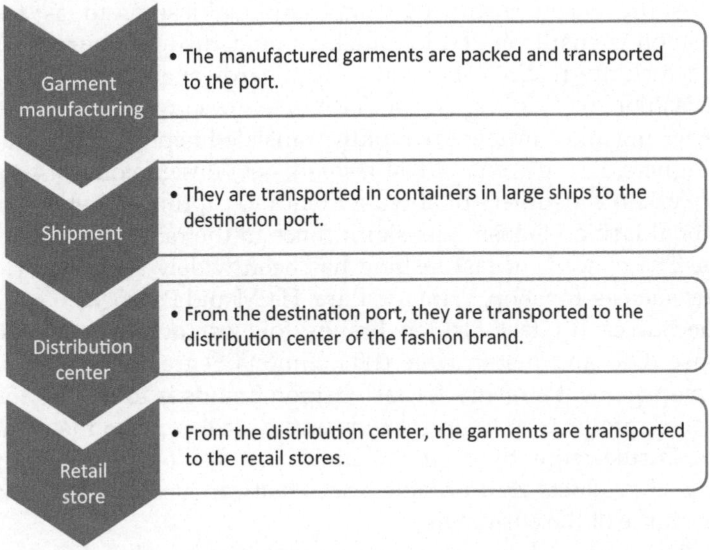

Figure 6.1 The logistics process of fashion garments.

as Walmart (USA), Marks & Spencer (UK) and Metro Group (Germany) to adopt the technology in their supply chain (Kelly and Erickson, 2005). Since then many other retailers from all around the world (such as Wesfarmers, Australia and Tesco, UK) have started implementing RFID technology into their supply chain network. Now the application of RFID technology is increasing at a very fast rate in diversified applications.

The retailers in developed countries in the USA and Europe are the leaders in the application of RFID into their supply chain network (Shin and Eksioglu, 2015). In developing countries, the RFID is an emerging technology to be adopted by many fashion retailers in supply chain and manufacturing. Hence, it creates lots of research potential in many retail industries including fashion. Although there are certain limitations of RFID technology, still the future of the RFID is very bright in fashion logistics.

Fashion logistics can benefit in several ways from adopting RFID technology (Azevedo and Ferreira, 2009; Bottani, 2009). Fashion logistics processes are rather complex involving many upstream and downstream suppliers including fiber, yarn, fabric, garment, trims (such as buttons, zippers, labels, etc.) and accessories (paper, plastic, metal and other packaging material). Problems encountered by any of the suppliers can impact the whole cycle of supply chain. The implementation of RFID technology can help in identifying the bottlenecks among suppliers, which can be useful for finding solutions or alternative approaches.

Consumers' taste is shifting from traditional fashion to fast fashion. Fast fashion is characterized by quick manufacturing of current trendy styles, which are retailed inexpensively (Bhardwaj and Fairhurst, 2010). In fast fashion, new concepts from fashion-shows organized in autumn, winter, spring and summer are rapidly translated to physical products by the designers. In the arena of fast fashion, new design concepts go from the catwalk to consumers in about a month to capture target customers. Traditional fashion is losing its significance in the race with fast fashion brands. The concept of fast fashion has been widely adopted by many retailers such as Topshop, Primark, Zara, H&M and Peacocks.

The success of fast fashion brands follows the concept of Quick Response (QR) and Just-in-Time (JIT) (Simona Segre, 2005). The supply chain management strategy for fast fashion brands is to deliver the current trendy products at the right time at the right price to the right consumers. Furthermore, the supply chain of fast fashion focuses on wider varieties of products at affordable prices that can also be customized as per the choice of the customers.

To be successful in the global competitive marketplace, fashion brands need to reduce their lead time, quickly respond to changes in fashion style, adopt automatic tools in the logistics process and improve information flow both in the upstream and downstream of the supply chain. In many instances the lack of information flow causes bottlenecks and delays in the supply chain process. To avoid this, fashion brands are trying to establish good vertical and horizontal relationships with all their upstream and downstream networks (Castelli and Brun, 2010).

RFID technology can improve the accuracy of information available during supply chain activities at any point, which is the key challenge in effective supply chain management. The following points describe the benefits of using RFID in fashion supply chain.

- **High product varieties:** Fashion products are characterized by high product varieties, which are difficult to uniquely identify and separate from a lot (Hinkka et al., 2015). The varieties include size variations, style variations, color and material variations. The use of RFID technology can uniquely identify each garment.
- **Short fashion cycle:** As the recent trend is moving towards fast fashion, the fashion cycle is getting shorter. Hence, keeping accurate sales or stock data is essential. RFID tags can provide information on fast-moving and slow-moving products. This will help merchandising managers to decide which products to replenish and which to mark down.
- **On-time manufacturing:** On the manufacturing floor, the use of RFID tags can provide real-time data on work-in-progress (WIP).

This data can be utilized to streamline slower processes on the shop floor and deliver the product within the lead time.

- **On-time arrival at retail:** From a retailer's perspective, as the fashion cycle is very short, the on-time arrival of merchandise is crucial (Bottani, 2009). The ability of RFID tags to track and trace fashion merchandise can streamline many activities along the supply chain for on-time arrival.
- **Increased visibility:** RFID technology can retrieve accurate information from upstream suppliers. If the tags are attached at the early stage of manufacturing, other members in the supply chain can also use the information to update their product catalogues and brochures further in advance (Hinkka et al., 2015). The information can also be used to avoid any confusion or error in the supply chain.
- **Increased efficiency and accuracy:** The application of RFID eliminates the manual process of scanning, which increases efficiency. Similarly, RFID technology reduces the chances of shrinkage, hence, the accuracy of the inventory.

6.2 *Scope of RFID in Logistics*

In supply chain management, the RFID systems can be used for material handling, tracking of material flow, warehouse management, inventory management, stock replenishment, prevention of shrinkage and improved management of assets (Nayak et al., 2015). RFID is a technology that helps business to identify specific products and all its related information as well as to track them along the logistics network from the manufacturing point to the point-of-sale (Dai and Tseng, 2011). Many industries and retailers are adapting RFID technology in their logistics activities because of the multifaceted benefits achieved as discussed above.

The major advantages of RFID technology over the barcode system is the elimination of the longer process of manual scanning and a quicker process for identification and documentation. Although the barcode system is comparatively cheaper, it needs manual scanning, and is slow and intermittent. On the other hand, the RFID system can complete the work in very short time and in a continuous process. RFID technology has the potential to read and collect information from multiple sources simultaneously without any scanning, which is the key benefit in the logistics process. This can help companies to gain competitive advantages by effective control and monitoring of inventory. The goals of RFID applications in supply chain of fashion include (but are not limited to):

- Minimizing overstock or understock situations.
- Identifying faster selling items for reordering and slower selling items for notifying products that remain on the shelf for too long.

- Simultaneously checking and tracking multiple products at remote locations.
- Updating real-time information through the computer system in warehouses and stores, reducing inventory inaccuracies.

The developments in electronic data interchange (EDI), electronic product code (EPC) and information technology have helped to broaden the application field of RFID technology.

6.3 Application of RFID in fashion logistics

In fashion logistics, RFID technology can be used in many operations such as warehousing, tracing and tracking of products, inventory management, reducing shrinkage and self-replenishment of stock, which are discussed in the following sections.

6.3.1 Warehouse management

A warehouse in a fashion supply chain receives goods, verifies the physical quantity, stores goods or performs cross-docking, performs value added activities (if needed) and issues goods for transportation to customers (Bottani et al., 2009). In the fashion supply chain, the warehousing operations are used by the manufacturers, logistics service providers (LSPs) and the retailers. The manufacturers store the raw materials and the finished products in their warehouse. The LSPs use their warehouses or warehouse services available close to the port for temporary storage (Bolumole, 2001). The retailers use their warehouses or distribution centers for collections storage and distribution of the fashion products. In all these warehouses, the RFID technology is increasingly used to automate the warehouse operations by eliminating manual work. The objectives of using RFID in the warehouses are:

- To minimize the cost of warehousing the fashion inventory;
- To reduce the time taken to inspect, store and issue goods from the warehouse;
- To increase the efficiency of using warehouse space; and
- To improve customer service by increasing efficiency and accuracy.

As the cost of RFID tags gradually decrease due to wider application, RFID technology will become an integral part of fashion warehousing (Ustundag and Tanyas, 2009). As many manual operations are replaced with the automation achieved by warehousing, the cost of labor will be further reduced. In an economy of increasing labor cost, RFID can reduce the cost of operation. Furthermore, the accuracy of physical volume and location identification increases with RFID, which can reduce the time to rework or locate goods in the warehouse.

RFID technology significantly reduces processing time, which can reduce the overall throughput time. The information contained in RFID tags can be retrieved at any point in the warehouse; the decision-making process in the fashion supply chain becomes simpler. The use of RFID technology improves the accuracy of the physical data and location of goods and movable assets (such as operators, forklifts and containers). This data can be integrated with the planning software, which can help in improving inventory planning and asset utilization.

The following section describes various warehouse operations and the application of RFID technology.

6.3.1.1 Order receiving and sorting

The receiving process includes the arrival of the goods at the warehouse. The warehouse associated with the fashion supply chain deals with products such as fabrics, trims and accessories. Once these goods arrive at the warehouse, they are verified against the bill of lading for their type and quantity. Inaccuracies in physical quantity or damage of the products (if any) are communicated to the supplier and the information is updated in the warehouse management system (WMS) (Yan et al., 2008). Following inspection, the goods need to be sorted according to their type and destination. For easy identification, a stock keeping unit (SKU) is assigned to the goods received, then the goods are carried out to designated places in the warehouse and stored.

The traditional process of performing the above-mentioned activities includes manual work, which is labor intensive and time consuming. However, the use of RFID technology can make the process much simpler. The physical quantity of the goods received and the product details are automatically updated from the RFID tags. The process of scanning individual barcodes is eliminated, which saves time and improves accuracy and efficiency.

The manual process of retagging with a separate barcode for identification within the warehouse and the SKU number allocation is also eliminated. The information now can be easily processed into the RFID tags. Furthermore, any lot received with damaged goods can also be easily identified with the RFID tags. During the incoming inspection, the damaged lot information is updated into the RFID tags, which can be used whenever necessary during the supply chain process.

6.3.1.2 Storage

Once the SKUs are allocated to the goods, the goods are carried to their storage places either by an automatic material handling system or manually. In the case of fashion warehousing, fashion goods such as coats and jackets need to be transported in hangers. Hence, many of the warehouses have installed hanger conveyor systems, which can be used for carrying the goods to their designated places where they are stored on the hangers (Choy et al., 2017). Many fashion goods are stored in the boxes as received, which can be managed by automatic conveyor systems.

The use of RFID technology can streamline the storage process by easily identifiable location and storage conditions. The person involved in the storage or the forklift operator can get the information from the RFID tags and accurately store the products. The RFID tags are automatically read when they enter the storage place and inventory is updated.

6.3.1.3 *Cross-docking*

In logistics, cross-docking is the practice of unloading incoming consignments from an incoming trailer, truck or railroad container and reloading them directly into the outbound trailers, trucks or rail cars, with little or no storage in between (Chow et al., 2006). This may be done to change the type of conveyance, to sort material intended for different destinations or to combine material from different origins into transport vehicles (or containers) with the same or similar destinations. Cross-docking is an important part in the fashion logistics, as it saves time, reduces the amount of material handling and consumers receive the product on time.

The use of RFID technology can increase the accuracy and efficiency of the cross-docking process. Once the fashion goods are received at the warehouse, the information is updated automatically, and their destination can be easily identified from the RFID tags. Hence, these goods are prepared to be transported to their destination by suitable transportation system without any storage in the warehouse. Without the RFID system, the received goods required manual checking and identification of their destination, which is a time-consuming process. Furthermore, the goods received and sent directly by cross-docking are automatically updated in the warehouse management system.

6.3.1.4 *Order picking*

Order picking is the process of selecting the right product in the right quantity for delivering to the customer. This step is crucial in warehouse management when dealing with a large variety of stock. For many retailers, full-pallet delivery has been replaced with mixed SKUs in the same pallet. Hence, identifying each SKU and its delivery is a time-consuming process. Any mistake in the selection of the right type of product can lead to customer dissatisfaction, return of the order and financial loss (Poon et al., 2009). The use of RFID technology can increase the efficiency and accuracy in order picking and packaging. In addition, RFID systems can simultaneously handle multiple information compared to the barcode system. Moreover, barcode systems can only handle one barcode at a time in a stop-go process.

6.3.1.5 Managing warehouse assets

Modern warehouses are equipped with equipment such as pallets, packs or lifters, lifter transporters, hand trucks, shrink wrap equipment, trailers, trucks and forklifts (Delen et al., 2007). Managing a large amount of equipment in warehouses is rather a difficult task. Many times, some of the equipment are kept in locations, which are hard to find. Improper planning also leads to equipment shortages at certain times and excessive equipment at others. Hence, the process of transportation and storage is delayed and there is time wastage in locating the equipment.

The use of high frequency RFID systems can help to easily trace the warehouse equipment. When the equipment moves from one place to another, the system updates the information from the RFID tags. At any point, the number of pieces of equipment sitting idle can be easily calculated. This can help the warehouse operations manager to keep accurate data on the utilization of each piece of equipment. From this data, the percentage utilization or hours of operation of each piece of equipment can be calculated, which can be used for the maintenance of the equipment and for the planning and future requirements of specific types of equipment. Furthermore, this data can also be used to improve the percentage utilization of expensive equipment, and waiting time of goods can be reduced.

6.3.2 Loading and unloading of cargo

Many of the fashion manufacturing industries are located in developing countries such as Vietnam, Bangladesh, India, Laos, China, Sri Lanka and Cambodia. The developed countries like the USA, Europe and Australia depend on the developing countries for clothing, textiles and home furnishings. Traditional fashion goods are generally transported via ship from the manufacturing countries to the consumers in developed countries. However, many fast fashion brands use air transportation for the rapid transport of goods. In addition, local transportation is done by rail and trucks. Hence, all the fashion brands use an intermodal transportation system.

Fashion manufacturers depend on third-party logistic (TPL) providers such as DHL, UPS, Ryder and FedEx for logistics activities. The TPLs are mainly involved in the collection, storage, documentation, transportation, value-addition and distribution of the fashion goods (Hsu et al., 2009). The major objectives of the TPL providers are to deliver the right product, to the right customer, at the right time, at the right price and right quality. Hence, accurate information flow is as essential step in the material flows. Generally, the information on the materials loaded into the ships is verified to reduce inaccuracies. The verification process is an additional step, which needs time, people and space. This process increases the logistics cost.

The use of RFID technology makes the loading and unloading of the cargo from ship much simpler. For example, fashion goods with RFID tags will be automatically scanned when the goods enter or leave the ship. Hence, the additional process of verification is eliminated. Unlike a barcode scanner, the physical movement and amount of material loaded into or removed from the cargo can be automatically monitored, which saves time and increases efficiency and accuracy. Similarly, RFID tags can be used to track material movement within or away from the ship, which increases visibility, improves coordination in the fashion supply chain and reduces shrinkage (Lai et al., 2005).

6.3.3 *Inventory management*

Inventory is the material stock an organization needs to meet its business objectives. Inventory in a fashion business includes the raw materials (yarns, fabrics, trims and accessories), in-process goods (such as fabric patterns in stitching) and finished garments ready for shipment.

For garment manufacturing industries, the major inventories are the fabrics, trims and accessories. Many of the industries deal with a wide range of woven and knitted fabrics with varying construction, weight and color. Similarly, the trims such as buttons, zippers, threads and accessories needed for final packaging vary in their color, size and properties. Managing such a wide range of inventory is rather a difficult task.

RFID tags can make the complex inventory management process much simpler for manufacturing industries (Nayak et al., 2007). Each fabric batch slightly varies in color from the others. Hence, the use of RFID tags can help to store the fabrics as per their batch. Furthermore, the information about the storage location and the yardage can be easily identified from the RFID tags. These processes are still managed manually in many industries.

Inventory is the important asset of a business that is ready or will be ready for sale. Inventory management is a discipline primarily about specifying the shape and placement of stocked goods. It is required at different locations within a facility or within many locations of a supply network to precede the regular and planned course of production and stock of materials.

6.3.4 *Out of shelf but in stock – shelf replenishment*

There are situations where items are out of shelf but still available in stock, leading to a decline of customer satisfaction and missed purchase

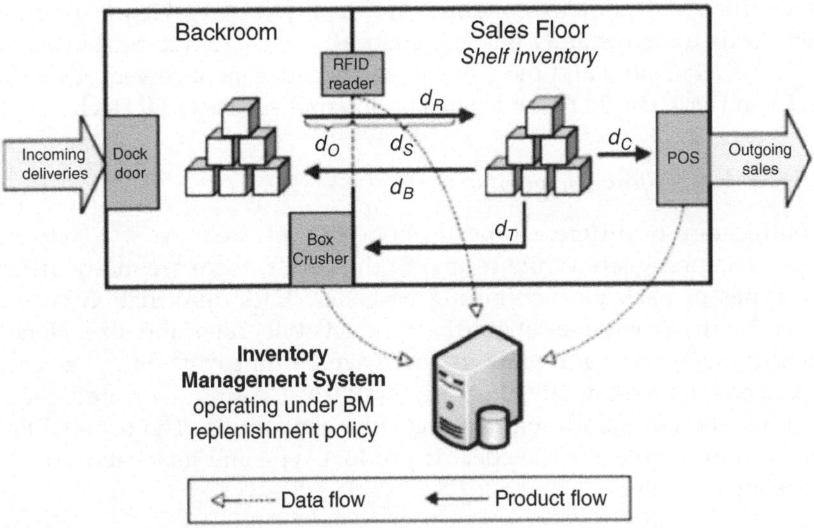

Figure 6.2 Inventory issues in the logistics process.

Source: Condea et al, 2012.

opportunities. Application of RFID in inventory management can reduce these issues and improve overall outcomes. By using the trace history, the stock available both on the shelf in the retail store and in the warehouse can be calculated and updated in real time, providing a more visible picture to the staff about stock levels (Zhu et al., 2012). The linkage between the store inventory and the warehouse inventory is shown in Figure 6.2.

6.3.5 *Out-of-stock replenishment*

Another benefit of RFID is triggering replenishment or reordering for any particular item that is out-of-stock or low in inventory. As mentioned before, RFID provides a clear and visible insight of the real-time stock levels both on shelf and in inventory, thus it can generate replenishment when necessary. RFID can help in avoiding a no-sale issue when a product is listed as in stock but is actually physically out-of-stock due to inventory levels not being properly maintained (Al-Kassab et al., 2013).

6.3.6 *Return/exchanged goods inventory control*

Returned items could complicate and disturb supply chain movement, especially in inventory management. Returned goods are often considered as a burden in relation to economic value and time spent. Hence, information needs to be instantly and automatically updated into the system of both the retail store and the warehouse for better management (De Brito and Van Der Laan, 2009), which can be accomplished by RFID tech.

6.3.7 *Identifying the goods*

Identification of different types of goods in a warehouse is a manual, time-consuming job. While receiving the goods, there are many different types of packages containing barcodes. Little information is provided by these barcodes about the product style, color and size. Hence, identifying specific garments in boxes can be time consuming, as each box needs to be manually checked. Similarly, locating the stored goods in a warehouse is both manual and time consuming. The use of RFID tags can exactly locate the specific product type and its location in the warehouse easily.

6.3.8 *Self-replenishment of stock*

RFID technology can monitor the amount of stock of any type of inventory in the warehouse. RFID tags help to accurately update the stock record when they are used or issued. When the stock level reaches the minimum threshold, an automatic warning system can alert the warehouse manager to initiate the reordering process, allowing the warehouse manager to start the ordering process for the timely replenishment of the stock.

6.3.9 *Reducing shrinkage*

Shrinkage is one of the problems faced by many fashion brands inside retail stores and warehouses (Tajima, 2007). The availability of readily manufactured garments in warehouses makes fashion goods vulnerable to theft, which is known as shrinkage. The barcode system needs the physical scanning of tags to collect information. Hence, goods with the barcode system can be stolen at any time, especially during night shifts when fewer employees are at work. The use of RFID tags can provide 24-hour protection, as they are active all the time. An alarm attached to an RFID system can alert security guards or managers in the warehouse when any unauthorized movement of goods is detected. Similarly, the RFID tags during transportation by trucks and even the shipment process can help to reduce the shrinkage. Any theft during transportation can be easily detected from the radio signals received from the RFID tags, where they were taken out from the storage. The chance of theft or shrinkage can be reduced as RFID tags increases the accuracy and visibility of products during storage and transportation.

6.3.10 Use of RFID in textile logistics

In many developed countries, fashion manufacturers import fibers, yarns and fabrics; where the quality of the domestic market does not meet the requirements of the international market. Similarly, the developed countries also export fibers, yarns and fabrics to other countries. The shipment process follows the same sequence as shipping the garments. However, the packaging requirements and the taxes and tariffs may be different than for finished garments depending on the nature of the products. Hence, RFID technology can also be used during the logistics activities of textiles. Various operations such as warehouse management, inventory management, tracing and tracking can be applied to the textiles during the logistics process.

6.4 RFID sample working process in logistics

Figure 6.3 illustrates logistics activities with the use of RFID technology. For example, after the manufacturer finishes all the processes of sewing, finishing and packaging of various garment sizes and types for a specific buyer, the goods need to be sorted and packed into boxes or palettes that

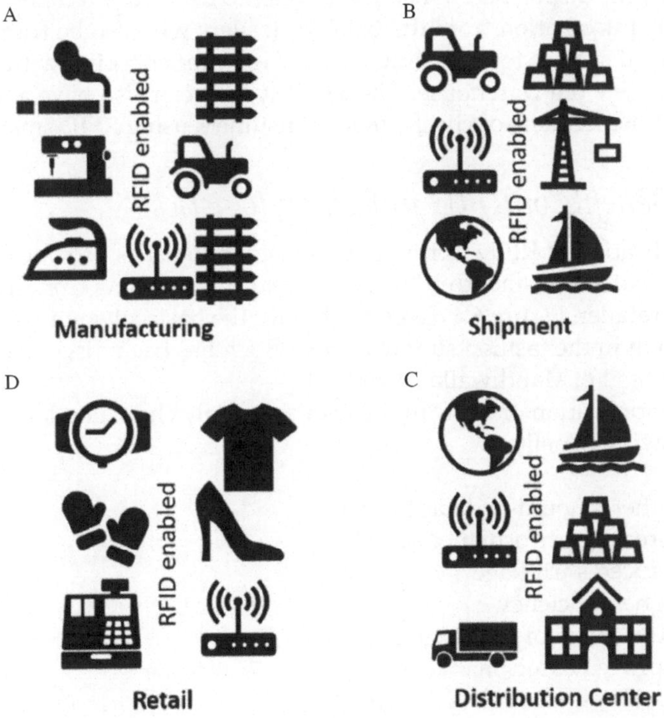

Figure 6.3 RFID sample in supply chain.

have been attached with RFID tags, making it easier to identify the boxes or pallets for a specific type of garment while packaging. Once, the garments are packed in the boxes or pallets, the system is updated. This saves the time of manually counting the stock and updating the system.

Those palettes are then delivered to port by trucks or trailers for shipping (Point A). At the loading dock, a reader quickly checks all fashion goods with RFID tags using "singulation method" and sends the information directly to the warehouse data management program. This program will uniquely receive and divide goods' data into separate locations of the system without any error and avoiding any congestion of the network. The information will be held on the interface for each type of product, manufacturer, destination and logistic providers, which is then transferred to the receiver end at the destination port. This data will reduce distributor time to check and identify items one more time (Point B).

When the fashion goods are received at the warehouse of the distributor, a similar process is performed to verify and separate goods, based on their type and destination, onto suitable trucks or into the warehouse to get ready for launch (Point C). The distribution center's warehouse data management program automatically updates the information on goods received and dispatched. Once the goods arrive at the retailer or store (Point D), information from the truck or trailers will also be transmitted to the retail system for receiving inventories after checking with readers located inside the warehouse. The RFID system can also have a hand on replenishment orders or check out for consumers using RFID smart cards.

6.5 *Benefits of RFID in fashion logistics*

The application of RFID technology in fashion logistics provides various benefits that allow smooth information and product flow from manufacturer to retailer. Figure 6.4 describes briefly the basic advantages of RFID application in the fashion supply chain, which has been discussed in this section (Assif & Mandiwalla 2005).

The application of RFID in the fashion supply chain and logistics has the following benefits:

- Higher amount of data storage
- Increased performance speed
- Reduced shrinkage
- Higher efficiency
- Better inventory management
- Improved assets management
- On time stock replenishments
- Real-time product availability
- Improved tracking of goods

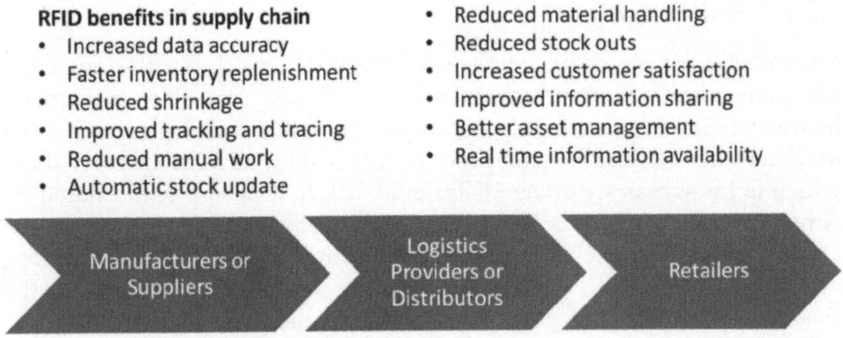

Figure 6.4 Benefits of RFID in supply chain.

6.5.1 *Higher data storage*

In term of data storage, RFID is a more advanced application than barcoding. An RFID tag in retail store can store up to approximately 2000 bytes of data. For example, one byte equals one typed character and 2000 bytes is equivalent to a short email text. Thus, the storage of one tag is enough to save all the related information of a product (Roberti, 2013).

6.5.2 *Increased performance speed*

In comparison to barcode technology, RFID is much faster in collecting data and not required to be within the visual line of sight of the reader; it can recognize the tag even from a long distance. Previously, with barcoding, staff needed to scan each product, then input the product details manually (Bragg, 2011).

6.5.3 *Utilize real-time information*

In the fashion industry, there are a wide range of product categories such as fabrics, buttons, zippers, labels and completed garments (not to mention there also many product SKU that change according to season); hence, it is very complex to manage all these components. RFID can manage this information easily by capturing updated information of large quantities of items simultaneously (Bragg, 2011). The product movement in the fashion supply chain is shown in Figure 6.5.

Figure 6.5 Product movement in the fashion supply chain.

6.5.4 Reduced shrinkage

The fashion supply chain involves a complex network of different organizations manufacturing various materials, trims, accessories and embellishments. During the supply chain process (shown in Figure 6.5) from manufacturers to retail, some products are lost by misplacement or theft, which is known as shrinkage (Zhu et al., 2012). It has been estimated by Advanced Market Research (AMR) that the shrinkage of inventory in the US has accounted for US$30 billion during the supply chain process (Garf and Langdoc, 2004). RFID tags attached to retail merchandise allows the system to recognize exactly a product's movement and its real-time location. Therefore, inventory loss and shrinkage will be avoided, helping the company to save time, cost and labor, since only one simple step needs to be done by scanning the tag, and no manual inventory checking or inventory movement is required.

6.5.5 Reducing repetitive manual task

RFID demonstrates an excellent usage in the cross-docking and tracking goods process. Traditionally, when receiving goods from manufacturers, the pallets or hangers had to be counted and checked manually to ensure the right quantity and the right product even though these processes had been performed several times before while in the production, packaging and delivering stages. RFID eliminates these steps by quickly scanning through the tags and readers; all items can be counted in just seconds. This is because RFID readers can read and capture all tags at the same time, even without being in the visible sight of the readers. For a business, RFID helps to save time and operation cost by reducing human errors and repetitive manual tasks, which may causes inaccuracies and delays (Wen, 2010).

6.6 Conclusion

Fashion logistics activities are undergoing a major change from manual methods to automated methods. Many fashion manufacturers transport their fashion garments using third-party logistics (TPL) service providers. The logistics providers in developing countries generally use traditional methods of record keeping, identification, transportation and warehousing, which is time consuming and inaccurate. The use of RFID technology can help in accurate information on inventory status and automatic stock updates. Many fashion brands are mandating the use of RFID technology by garment manufacturers, whereas some of them are implementing this technology after seeing their competitors benefit from it. Although the cost of implementing RFID technology is high, the return on investment can be realized in the long run. In future, RFID will be adopted by almost all logistics activities due to improved technology, cost reduction and subsidies by government to implement the technology.

References

Al-Kassab, J., Thiesse, F. & Buckel, T. 2013. RFID-enabled business process intelligence in retail stores: a case report. *Journal of Theoretical and Applied Electronic Commerce Research*, 8, 112-137.

Asif, Z. & Mandviwalla, M. 2005. Integrating the supply chain with RFID: A technical and business analysis. *Communications of the Association for Information Systems*, 15, 24.

Asif, Z. & Munir, M. 2005. Integrating the supply chain with RFID: A technical and business analysis. *Communications of the Association for Information Systems*, 15, 393-426.

Azevedo, S. G. & Ferreira, J. 2009. RFID technology in retailing: an exploratory study on fashion apparels. IUP *Journal of Managerial Economics*, 7, 7-22.

Bhardwaj, V. & Fairhurst, A. 2010. Fast fashion: response to changes in the fashion industry. *The International Review of Retail, Distribution and Consumer Research*, 20, 165-173.

Bolumole, Y. A. 2001. The supply chain role of third-party logistics providers. *The International Journal of Logistics Management*, 12, 87-102.

Bottani, E., Ferretti, G., Montanari, R. & Rizzi, A. 2009. The impact of RFID technology on logistics processes of the fashion industry supply chain. *International Journal of RF Technologies: Research and Applications*, 1, 225-252.

Bragg, S. M. 2011. Inventory best practices, John Wiley & Sons.

Castelli, C. M. & Brun, A. 2010. Alignment of retail channels in the fashion supply chain: An empirical study of Italian fashion retailers. *International Journal of Retail & Distribution Management*, 38, 24-44.

Chow, H. K., Choy, K. L., Lee, W. & Lau, K. 2006. Design of a RFID case-based resource management system for warehouse operations. *Expert Systems with Applications*, 30, 561-576.

Choy, K. L., Ho, G. T. & Lee, C. 2017. A RFID-based storage assignment system for enhancing the efficiency of order picking. *Journal of Intelligent Manufacturing*, 28, 111-129.

Christopher, M., Lowson, R. & Peck, H. 2004. Creating agile supply chains in the fashion industry. *International Journal of Retail & Distribution Management*, 32, 367-376.

Christopher, M. & Peck, H. 1997. Managing logistics in fashion markets. *The International Journal of Logistics Management*, 8, 63-74.

Condea, C., Thiesse, F. & Fleisch, E. 2012. RFID-enabled shelf replenishment with backroom monitoring in retail stores. *Decision Support Systems*, 52, 839-849.

Dai, H. & Tseng, M. M. 2011. Determination of production lot size and DC location in manufacturer? DC? retailer supply chains. *International Journal of Logistics Systems and Management*, 8, 284-297.

De Brito, M. P. & Van Der Laan, E. A. 2009. Inventory control with product returns: The impact of imperfect information. *European Journal of Operational Research*, 194, 85-101.

Delen, D., Hardgrave, B. C. & Sharda, R. 2007. RFID for better supply-chain management through enhanced information visibility. *Production and Operations Management*, 16, 613-624.

Garg, R. & Langdoc, S. 2004. *Shrink reduction strategies – proactively control lost inventory* [Online]. Available: http://securitysource.com/docs/amr1.pdf [Accessed 18 March 2018].

Hinkka, V., Häkkinen, M., Holmström, J. & Främling, K. 2015. Supply chain typology for configuring cost-efficient tracking in fashion logistics. *The International Journal of Logistics Management*, 26, 42-60.

Hsu, C.-I., Shih, H.-H. & Wang, W.-C. 2009. Applying RFID to reduce delay in import cargo customs clearance process. *Computers & Industrial Engineering*, 57, 506-519.

Kelly, E. P. & Erickson, G. S. 2005. RFID tags: commercial applications vs. privacy rights. *Industrial Management & Data Systems*, 105, 703-713.

Lai, F., Hutchinson, J. & Zhang, G. 2005. Radio frequency identification (RFID) in China: opportunities and challenges. *International Journal of Retail & Distribution Management*, 33, 905-916.

Masson, R., Iosif, L., Mackerron, G. & Fernie, J. 2007. Managing complexity in agile global fashion industry supply chains. *The International Journal of Logistics Management*, 18, 238-254.

Nayak, R., Chatterjee, K., Khurana, G. & Khandual, A. 2007. RFID: Tagging the new era. *Man-Made Textiles in India*, 50, 174-177.

Nayak, R. & Padhye, R. 2014. Introduction: the apparel industry. *In:* Nayak, R. & Padhye, R. (eds.) *Garment Manufacturing Technology*. Elsevier.

Nayak, R. & Padhye, R. 2015. *Garment Manufacturing Technology*, Cambridge, UK, Woodhead Publishing.

Nayak, R. & Padhye, R. 2017. In *Automation in Garment Manufacturing*, Woodhead Publishing.

Nayak, R. & Padhye, R. 2018a. Artificial intelligence and its application in the apparel industry. *In:* Nayak, R. & Padhye, R. (eds.) *Automation in Garment Manufacturing*. Elsevier.

Nayak, R. & Padhye, R. 2018b. Introduction to automation in garment manufacturing. *In:* Nayak, R. & Padhye, R. (eds.) *Automation in Garment Manufacturing*. Elsevier.

Nayak, R., Singh, A., Padhye, R. & Wang, L. 2015. RFID in textile and clothing manufacturing: technology and challenges. *Fashion and Textiles*, 2, 1-16.

Poon, T., Choy, K. L., Chow, H. K., Lau, H. C., Chan, F. T. & Ho, K. 2009. A RFID case-based logistics resource management system for managing order-picking operations in warehouses. *Expert Systems with Applications*, 36, 8277-8301.

Roberti, M. 2013. Storefront Backtalk gets its RFID facts wrong—again [Online]. Available: http://www.rfidjournal.com/blogs/rfid-journal/entry?10484 [Accessed 11 July 2018].

Shin, S. & Eksioglu, B. 2015. An empirical study of RFID productivity in the US retail supply chain. *International Journal of Production Economics*, 163, 89-96.

Simona Segre, R. 2005. China and Italy: fast fashion versus Pret a Porter. Towards a new culture of fashion. *Fashion Theory*, 9, 43-56.

Tajima, M. 2007. Strategic value of RFID in supply chain management. *Journal of Purchasing and Supply Management*, 13, 261-273.

Ustundag, A. & Tanyas, M. 2009. The impacts of radio frequency identification (RFID) technology on supply chain costs. *Transportation Research Part E: Logistics and Transportation Review*, 45, 29-38.

Wen, W. 2010. An intelligent traffic management expert system with RFID technology. *Expert Systems with Applications*, 37, 3024-3035.

Yan, B., Chen, Y. & Meng, X. RFID technology applied in warehouse management system. Computing, International Colloquium on Communication, Control, and Management, 2008. ICCCM'08. 2008. IEEE, 363-367.

Zhu, X., Mukhopadhyay, S. K. & Kurata, H. 2012. A review of RFID technology and its managerial applications in different industries. *Journal of Engineering and Technology Management*, 29, 152-167.

chapter seven

Application of RFID in fashion retail

KEYWORDS: *retail management, merchandise replenishment, customer relationship management, fitting room, shrinkage reduction*

7.1 Introduction

Fashion retail is the marketing place for consumers to purchase their fashion goods via various channels of distribution (Barnes and Lea-Greenwood, 2010). Retailers try to fulfill customer demand through various upstream suppliers and manufacturers such as fiber manufacturers, yarn and fabric manufacturers and processing industries. In fashion retail the demand of a large customer range is fulfilled with fashion products, which are volatile and varying in nature. Fashion retail stores may deal with all or some fashion goods for kids, menswear, women's wear, seasonal wear and fashion accessories (Moon and Ngai, 2008; Nayak and Padhye, 2015b).

Contemporary fashion retail stores have undergone several changes due to the technological revolution, change in consumers' attitude, emergence of fast fashion and the influence of globalization (Gaukler et al., 2007). Furthermore, many fashion retail stores are either closing their physical retail outlets or adopting online selling due to the increased use of ecommerce. Although the price of raw materials and labor cost has been on the rise, consumers always want fashion goods at cheaper prices. On the other hand, the emergence of many new fashion brands has increased the global competitiveness for good quality fashion goods at cheaper prices (Battaglia et al., 2014).

The evolution of many new fashion brands, rise in ecommerce and the change in consumers' attitudes towards fast fashion have revolutionized the fashion retail sector (Bhardwaj and Fairhurst, 2010). Fashion retail stores face fierce completion from different brands including the rising online retail stores. Furthermore, advancements in technology and the supply chain process have changed the features of retail from what was normal two decades ago. Considered from an integrated approach, fashion retail covers the whole process of manufacturing, logistics, marketing and selling of products in-store and through digital channels.

The success of modern retail stores depends on the selection of the right product type as per consumers' choice at the right price and at the right time. Although it looks simple, retail management is a difficult concept to execute, especially in fashion retailing where the consumers' choice is highly volatile and the demand cycle is short (Stratton and Warburton, 2003). Overstocked retail inventories can lead to reduced return on investment (ROI), whereas understocked inventories can lead to lost sales and reduced consumers. Hence, retail stores should find various ways of effectively managing their inventory.

Modern retail stores must focus on strategic planning, customer service, competitive cost and new technologies such as Radio Frequency Identification (RFID) to be successful in the global competitive supply chain (Nayak et al., 2015). RFID technology is an emerging innovation that many fashion retailers are adopting and it is the prime investment for many fashion brands. It is estimated that more than 50% of retail business in developed countries will adopt this technology by 2020 (Azevedo and Ferreira, 2009). It has the potential to revolutionize the retail industry, which is currently dominated by the barcode system today. The major strength of RFID technology lies in the accuracy of product information and product tracking, which avoid the cumbersome manual counting method.

Accuracy of information helps to keep exact records of stock, stock replenishment and decision-making on slow-moving fashion goods for markdown purposes (Rekik et al., 2009). Product tracking without physical counting using an RFID system helps to locate fashion goods even if they are at the bottom of a pelleted container in a retail storage space or hiding some place within the retail store. The traditional process of barcoding for stock consolidation involves scanning of all the products, which is eliminated with the RFID system. In addition, the RFID system can indicate the type of items that are selling faster or slower, which provides customized information for increasing the sales.

Due to the advantages of the RFID system over the barcode system, several retail operations such as Walmart in the US, Tesco in the UK, Carrefour in France and Metro in Germany have already adopted the technology in their retail stores (Shin and Eksioglu, 2015). Similarly, fashion brands such as Tommy, Adidas, Nike, Zara and H&M are the frontrunners in adopting this technology. By the end of 2025, it is assumed that all the leading fashion brands will be using this technology in their whole supply chain process including at the retail end (Nayak et al., 2007). This chapter discusses various applications of RFID technology in fashion retail. RFID technology can be used for product tracking, retail management, merchandise replenishment, reducing shrinkage, brand segregation, in the fitting room, customer relationship management, counterfeit prevention and managing returns.

7.2 Description of fashion retail

Fashion retail operates under similar principles to other types of retail stores. In fashion retail, fashion brands manufacture products mainly in developing countries such as China, Vietnam, India and Bangladesh (Nayak and Padhye, 2015b; Nayak and Padhye, 2014b). These products are shipped to the warehouses of the fashion brands located in some centralized locations. The warehouses stock the products and transport them to retail stores when needed. Some warehouses work as distribution centers when there is no distribution center at other locations.

While receiving the products in the warehouses of the retail stores, the physical quantities, quality of products and product styles are verified. Any mismatch in the quantity or style or damaged products are identified or returned to the manufacturers. This information, including the total new stock, is updated into the database management system of the warehouse. The warehouse serves as a central place for the brand's retail stores as well as other distribution centers (DCs) operated by the brand.

The distribution centers send their orders to the warehouses, whereas retail stores send their orders either to the distribution centers or to the warehouses to replenish their stock. Upon receiving the order, the warehouses need to send the products within the minimum time. Hence, the employees at the warehouse collect the stock (exact quantity and style), fulfil the demand and update the stock record. Once the order is complete, it is loaded on to the appropriate transportation vehicle and then transported to the retail stores. The warehouses then work to replenish the stock from the manufacturer. The retail operations from garment manufacturers to consumers are described in Figure 7.1.

Once the goods are received at the retail store, they go to the storage area of the retail store. The packages are opened, and products are sent to be displayed in the shelves or on hangers in the retail store. The additional stock is held in the storage area for some days or even weeks. The retail

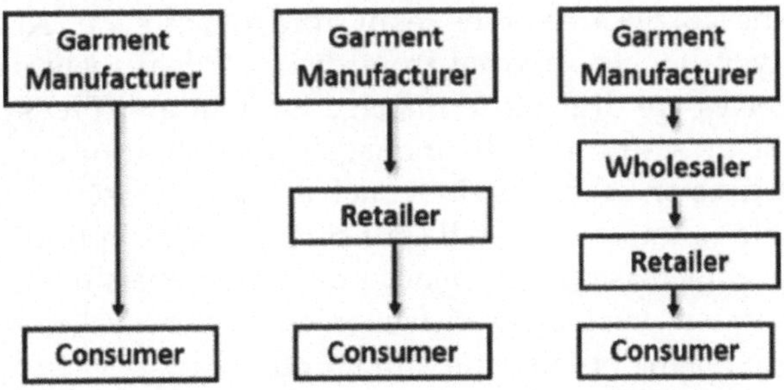

Figure 7.1 Fashion retail operation network from manufacturing.

stores need good display styles and visual merchandising of products for sales promotion. When a new product has been launched, it is displayed at the new arrival section. The customers visit the retail stores, select their products and buy them from the retail stores. RFID technology has many roles in the retail business to reduce manual work, keep the stock and automatically update stock information.

7.3 Application of RFID in retail

There are several applications of RFID technology in retail stores, which includes product tracking, inventory management, retail management, customer relationship management, brand segregation and reducing shrinkage. The major advantages are related to the easy tracking of inventory and on-time information availability, which increases the accuracy and visibility of the products in fashion retail.

7.3.1 Product tracking

In retail stores, inventory moves several times from the shelves to the fitting room or other locations, which often leads to false out-of-stock situations. This can lead to lost sales if the product is not tracked in a timely fashion and restored to the original shelves. It has been shown that about 65% of traditional inventories provide wrong information, as manual methods are used for inventory management. The use of RFID technology can provide solutions to these problems and produce accurate stock levels whenever needed (Gaukler et al., 2007).

The major advantage of RFID technology in product tracking is the clear visibility of products and accuracy of information (Brusey and McFarlane, 2009). Even the product hiding under the rack or buried under a pile of other products can be easily located using RFID tags. This in turn reduces the manual labor involved in tracking the product, improves the efficiency of shelf management and as a result net sales are increased. Without the RFID system, staff must spend a lot more time searching each product in the storage area of the retail store to locate the SKUs.

Similarly, products removed from their original shelves and placed on other shelves are difficult to manage without an RFID system. With an RFID system, staff can easily manage product tracking and therefore spend more time on other activities such as customer service. The use of RFID can help to improve overall staff service levels, optimize inventory utilization and increase the efficiency and effectiveness of stock management. Furthermore, customer satisfaction levels and loyalty are increased, which help in raising brand equity and profitability.

Intelligent Cloth-Racks and Smart Shelves can help retailers to gain additional intelligence about their customers' shopping and buying behaviors, which can then be used for further improving product and service offerings (Medeiros et al., 2008). Intelligent Cloth-Racks can be integrated with an RFID system. Each time a customer or employee removes a fashion item equipped with an RFID transponder from the rack, that movement can be automatically entered into the merchandise management system. The moment the article is returned can also be registered. The time that lapses before a customer returns an item to the rack provides valuable information. For example, if a customer puts a blouse back on the rack after five seconds, she possibly did not like the material. If five minutes went by, it is likely that the customer tried on the article. Smart Shelves can offer similar functionalities for stacked fashion goods.

7.3.2 Retail management

RFID technology can be used in retail stores to monitor and control the floor level out-of-stock situation (Roussos, 2006; Chen and Pfleuger Jr., 2008). It is mostly required where there is high product display density, low staffing level and a good chance of mishandling. With RFID technology, the stock level of items is properly maintained and items can be grouped according to their demand. The use of RFID technology improves the visibility of fashion goods and accessories, which can help in improved retail management.

While receiving goods at the retail stores from the distribution centers, the accuracy of the deliveries can be ensured quickly and with reduced labor (Twist, 2005). The fashion goods arrive in the retail store on pallets consisting of one or several styles of garments or other fashion accessories. Retailers mostly verify the accuracy and integrity of the received stock by manual open-box audits prior to stocking or storing of goods. But now it is possible to read all the items packed inside the carton without opening the boxes with the use of RFID technology, which saves time and labor cost.

The total stock can be automatically updated from the data collected by the RFID system, which avoids the time-consuming process of manual counting. This can improve the inventory accuracy and automatic replenishment of stock in the retail environment. The accuracy in stock can help to avoid out-of-stock situations and improve the availability of goods for sales. Therefore, the RFID provides a new horizon to retailers in the receiving and storage of materials, which eliminates invoice disputes and the mixing of different qualities and styles of products.

RFID has been used by a number of major retailers in Europe and the USA. Walmart is among the early adopters of RFID throughout their supply chain. Suppliers working with Walmart had been under pressure to implement RFID in their supply chain component as well. The reason for implementing RFID is to be in line with Walmart's corporate strategy of

Figure 7.2 An RFID tag attached to a high-end luxury product in a fashion retail store.

cost competitiveness and fast response (Vowels, 2006). After the successful implementation and use of RFID by Walmart, other big retailers like Marks & Spencer and Tesco in the UK are also adopting it, while the Coles group in Australia has given a thought to embracing this technology. Fashion industry is always regarded as a later adopter of new technologies than others, but in adopting RFID it is not lagging. An RFID tag attached to a high-end luxury product in a fashion retail store is shown in Figure 7.2.

7.3.3 Merchandise replenishment

An RFID system prevents brands from losing money by not having the right product on the shelf at the right time or avoids an out-of-stock situation by timely replenishment of stock (Condea et al., 2012). The retail store merchandise replenishment process presently used by many retail stores is completely manual, hence not very effective. It is time consuming and labor intensive and some popular items are not replenished for weeks. This can be overcome using RFID technology. The tags help to estimate stock levels frequently and there is an in-stock position improvement up to 30% (Nayak et al., 2015). This has a positive impact on net sales and customer satisfaction.

RFID technology in fashion retail can avoid having seasonal merchandise lie undetected in the back room and sold at significant markdowns, thus improving profit margins. If a specific product reaches the threshold stock level on a shelf, a notification will be sent to the stockroom to replenish the stock or to order more stock immediately. The RFID system can send the data to be automatically added into the new purchase order. In that way, products can be received just-in-time without excessive stock to fulfill consumer demand (Michael and McCathie, 2005). According to the 2017 Retail Vision Study, retailers being mentioned in the report have reduced the scenario of "out-of-stock" by 60 to 80 percent.

The use of RFID technology can help to convert the shelves in the retail environment into smart shelves for item-level picking (Medeiros et al., 2008). For example, when a consumer buys a specific product from a retail outlet, the RFID system can track the shelf that has the reduced stock due to consumer purchase. A replacement note will be sent to the retail stock manager, who can work with the replenishment staff to send the same item to the shelf. Furthermore, a complete list of the stock can be created by the RFID system of items that need to be collected from the retail storage area for replenishment in the sales floor.

These smart shelves can help to identify the number of days or weeks a product has been displayed without being sold. Seasonal fashion goods or festive fashion merchandise can also be identified using smart shelves. Retail merchandisers can decide when to apply markdowns for different styles that are not selling fast or for merchandise that is approaching end of season. This can help in enhancing retail traffic and hence gross profit.

The nature of stock replenishment in many retail stores is often complex, which is still done manually. Upgrading shelves to smart shelves can help retailers to reduce replenishment time, avoid out-of-stock situations, increase inventory accuracy, increase sales and finally increase retail profitability.

7.3.4 *Information to customers*

Today's customers of fashion goods want additional information relating to the product such as the type of material, country of manufacturing, the total carbon footprint and information relating to the triple bottom line (TBL) of sustainability. The use of RFID technology can help to provide the information to the customers. When the customers take the product to the information scanner, the RFID reader will collect the information relating to the product stored in the RFID tag and display it on the screen (Koh et al., 2006). Customers can know exactly the type of fiber used, the country of manufacturing and the use of ethical and sustainable practices during the manufacturing process. This information will help them to make proper decisions according to their choice and preferences. Customers are happy to know all this information and even pay higher prices for the cost of manufacturing that involves sustainable and ethical practices. This can increase the net sales of the retail store and its profitability.

7.3.5 *Reducing shrinkage*

Shrinkage is a major problem in many modern retail stores. For example, the total shrinkage loss in the retail sector for Australia was $4.5 billion and the US was $42 billion, respectively during 2017 (Huber and Michael, 2007). The total shrinkage was worth $100 billion worldwide, which was about 1.82% of total global sales. This shrinkage is a loss of the merchandise as a result of shoplifting, employee theft or other errors collectively. Shrinkage can also be related to inaccuracy in inventory, over counting of the inventory, improper stock conciliation, human errors during stock update, theft and errors in checkout (De Kok et al., 2008). In addition, the misplacement of clothes in the retail and incorrect price tagging cause shrinkage.

With the use of an RFID system in retail, each product can be closely monitored. With the RFID system enabled, the movement of the product is monitored. When consumers buy the product, the RFID tag is deactivated, and the operation is nullified (Dai and Tseng, 2012). Hence, the consumers can leave the store without any alarm buzzer. However, if the items are not paid for, the RFID system remains active. When consumers leave the store with the garment without paying, it can be detected, and the alarm alerts the security of the potential theft. Hence, shoplifting by consumers and the loss of products from the storage area can be reduced. Luxury fashion brands and jewelry stores are increasingly using the RFID system to avoid high financial loss due to shoplifting.

With the use of RFID technology, customers can be given advice on the fitting of the clothes they are trying on. This task can be achieved using smart try-out cabins and smart mirrors. The smart mirror can advise on additional clothing and fashion accessories as per the customer's preferences (Roberti, 2006). Global fashion retailers such as Marks & Spencer and Levi Strauss & Co. are installing RFID to acquire data on daily stock levels so that stock can be replenished to fulfill customer demand (Jones et al., 2005). However, without the RFID system in place, significant differences were observed between the data received from the manual systems and actual inventory levels (Jones et al., 2004).

Furthermore, at the point-of-sale the RFID tag can be read quickly and avoids the need for the physical handling of the product, as in a barcode system. RFID tags can be placed at the side seams or hem of a garment without affecting its aesthetic appeal. The number of items sold is automatically adjusted from the amount of stock. Hence, customers can be informed about the status/stock of merchandise by interactive display screens. The use of RFID tags has helped to reduce the out-of-stock rate at Walmart by 16%, and RFID-enabled stores were 63% more effective in replenishing out-of-stock items than were control stores (Age, 2005).

7.3.6 Theft detection in fitting room

The theft of garments from fitting rooms (or off the shelf) is a major problem in luxury and basic fashion retail stores (Melià-Seguí et al., 2013). Customers in fitting rooms can remove barcode tags and take the clothes with them when they leave the store. Fitting room theft can be prevented using an RFID system by mounting a small reader at the entry point of the fitting room. The tag information of the garment is captured when the customer takes the garment to the fitting room. Items taken to a fitting room but not returned are reported as potential loss. This can be reported to security via an alarm system. Thus, an RFID system can be used to identify the missing merchandise and to prosecute shoplifters (Rekik et al., 2008; Jaselskis et al., 1995). The provision of item-level tagging by RFID systems can also prevent theft by identifying the movement of certain items in the store. If the item is taken from the shelf and not returned, it can be reported as a loss if it is not detected in the store. If an item hasn't been checked and untagged at the payment counter and the customer leaves the store, retailers will be informed either by the store alert system or the RFID system (Michael & McCathie 2005).

7.3.7 RFID for brand segregation

RFID systems would be beneficial to department stores, which stock a range of products and brands under one roof (Loebbecke and Huyskens, 2008). The different brands can be controlled in terms of their inventory and location quite easily using RFID, whereas it would be quite a challenging task to separate any mixed brands without RFID tags. RFID also reduces human error while separating the brands, as this technology is not dependent on people to monitor inventory levels, identify brands or authenticate products.

RFID can also be used to avoid illegal product alteration for the grey market and to increase product security and information about customers' shopping behavior at retail outlets. With all these issues solved, textile and fashion supply chain can be made more streamlined and efficient in terms of cost and performance.

7.3.8 RFID used in fitting room

RFID technology is used in the fitting room in a new approach by a German department store—"Galeria Kaufhof in Essen"—which is a part of the Metro retailing group. Automatic suggestions are provided to customers buying clothes in the store when they go to a dressing room to try a suit (Melià-Seguí et al., 2013). A "smart mirror" informs them what kind of suit or accessories will be appropriate to them. An RFID reader on the smart mirror in the fitting room determines which clothing has been brought into the room from the RFID tag attached to the apparel. The appropriate accessories are then displayed on a suitable interface. This system is used in combination with "smart shelves", which can read what merchandise is currently in stock, so that customers can be shown various styles and sizes that are available.

The combination of RFID technology and various computing technologies can help to provide the feedback to the customers. The magic mirrors can be designed using the above technologies for providing suggestions to the consumers on their likings. In addition, the magic mirrors can also provide information on the price, promotions, availability of sizes and color, other items that match. A single touch on the mirrors can also help to bring the sales person into the fitting room for additional assistance. This would not only enrich customer experience in shopping but also boost the sales from a retail store and enhances the profit.

7.3.9 RFID tags for care labelling

The recent care labels are printed or woven labels storing a limited information (Nayak and Padhye, 2015a; Nayak and Padhye, 2014a). The electronic labels use RFID tags for storing information electronically on a garment. The same RFID tag used for containing the product details at the point-of-sale cannot be used for storing the wash care instructions, due to privacy concerns. Hence, additional tags are needed for the wash care instructions, which will increase the cost of the garment. In addition, consumers need special readers to extract the information stored in the RFID tag. Hence, customers prefer a physical label instead of an electronic label. Furthermore, the use of RFID tags on the garment may pose health risks to consumers due to exposure to radiation.

The other concern related to the use of RFID tag is electronic waste. An appropriate method should be devised to recycle or reuse RFID tags if they are removed at the point-of-sale. Similarly, if customers remove the chip at some point, proper disposal of the tags is essential, which can cause additional concerns for companies. In addition, other issues related to RFID are the lack of standardization and high cost. Standardizing the technology, design and use of tags can solve the issues related to lack of standardization, whereas technological developments can help to overcome cost-related issues.

RFID tags can be used in combination with the Registered Identification Number (RN) system, which is currently in use in some places including the USA. The Federal Trade Commission (FTC) issues the RN, which can be used on the textile label instead of the company name. The use of the RN system assists buyers to easily identify a company from the RN using the internet. The RN system uses less space on the label so more space is available for other care instructions. Therefore, the combined use of RFID and RN number systems could further benefit both industry and consumers.

The use of RFID tags is still in its infancy. Hence, it is too early to conclude the final shape of RFID technology and its impact on consumers. One company identified the benefit to consumers of including washing instructions on the RFID tag. In Germany, RFID chips are being used in nursing homes to help nurses correctly sort garments according to care instructions. Hence, there is a potential benefit of RFID tags to consumers. However, all the major drawbacks should be overcome to make RFID tags a success.

7.3.10 Customer relationship management

It is quite imperative to understand the needs and tastes of customers especially in the fashion industry, to offer them the right products. Moon and Ngai (2008) conducted a study on the use of RFID in fashion retailing and reflected that different customers behaved differently at different times in their shopping experience and knowing them in-depth regarding their shopping experience was very important so that sales personnel could provide them with personalized services. The participants in their study mentioned that if a customer entered the store and picked a few items connected to RFID technology, that would provide guidelines for their pattern of choices on that day so tailored advice could be provided while selecting the product.

When a customer enters a retail store, an RFID-enabled mobile app can give information to the customers on their choices. The app also can guide customers to the exact location where the product is available. Furthermore, customers can also receive the information on markdowns, promotions and special buys while finalizing their purchases. Hence, customers can get the maximum utilization for their time and money. Without an RFID system in place, customers will need the help of sales staff, move from shelf to shelf to find the products they want and not know about possible promotions. Hence, the use of RFID technology can help customers a lot in making the right decisions on their purchases.

As competition between fashion brands intensifies, it is important to satisfy customers by knowing their choices and providing updated product information (Uhrich et al., 2008). The RFID system can help to create a customer database, including their history of purchases, especially for high-end luxury products. Each time a specific customer buys a product, the RFID system can record and combine all the purchases made by the customer. A profile can be created that includes purchasing behavior, reward preferences, special interests and style preferences. This data can be used to provide professional suggestions for their next visit to the retail store. RFID can then be used to continually update information on target customers and their choices.

Since each customer's preferences are as different as their buying behaviors and buying patterns, RFID can be used to match products with consumers to suit their choices and assist them in making purchasing decisions. Subsequently, customers can be offered customized assistance to cater to their choices thus increasing revenues. Based on their purchasing behavior, customers can be updated with new arrivals, storewide events or any other relevant information to the customers. This can help them to make smart decisions and help the retail store to increase its sales.

Most fashion brands struggle with having a diversified customer profile for increasing sales through product promotion. An RFID system can be used to make a list of customers with the amount of money they have spent, their lifestyle and choices. Special loyalty cards can be provided to customers, which can inform them as to the availability of their preferred products. This will help the brand to create efficient and proper strategic planning for not only promotions like discounts or gifts but also visual merchandising in the store by analyzing consumer behavior. The application can be leveraged to mount and upgrade the target customer card loyalty program and the consumer-retailer relationship (Moon & Ngai, 2008).

7.3.11 Product accessibility in retail store

The use of RFID technology increases the visibility and accuracy of various fashion styles in the retail environment. Item-level tagging can reduce customer waiting time when requesting a certain product, since a salesperson is able to know the exact location of any item at any time. This could especially benefit multi-location retail stores in informing consumers of the exact store location that has the desired product. This is not only for retailer efficiency but also for improved consumer experience. This also increases sales potential, reduces manual work, reduces labor cost and improves inventory management.

7.3.12 Consumer self-checkout

Many of today's fashion retail stores are getting busier every day. This problem is aggravated during the festive seasons such as Christmas,

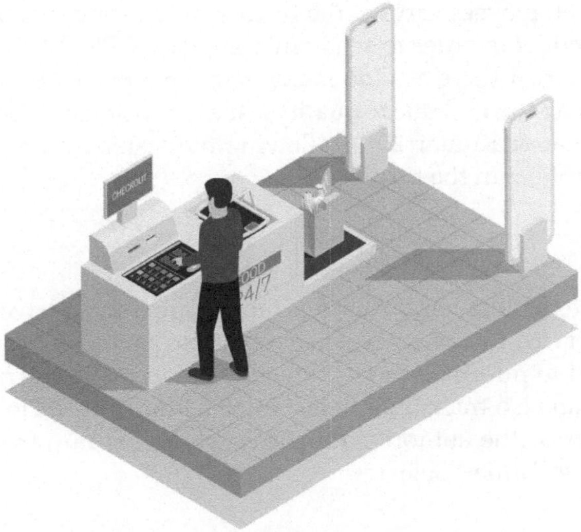

Figure 7.3 Self-checkout process with RFID system.

Source: ShutterStock.

Easter or New Year's celebrations. Customers need to stand in long queues to purchase products. This can be avoided using RFID technology, as RFID provides customers with the ability to check out all their products in one passage over the RFID scanner. Since the RFID system does not require line-of-sight access for scanning, the reader can detect all the products when they are passed over the scanning point. Customers can then check out with much less time and without the assistance from any internal staff (Reinhart et al., 2012). Figure 7.3 illustrates the process of self-checkout of customers with an RFID system where all items are being scanned at once.

An RFID system can also facilitate the payment process. After the fashion goods are scanned and the total payment has been calculated and displayed on the screen, the customers can swipe an RFID tag unique to the customer. This RFID tag stores the information relating to the customer's method of payment such as account details or credit card details. The purchase amount is then charged to the customer's personal account.

7.3.13 Managing returns

Managing returns or the reverse logistics issues can also be resolved by retrieving the information stored in the RFID tags. When the consumer returns certain goods, its RFID tag can be read and added to the database.

In that way, employees can see the history of who supplied or manufactured the product in order to send any complaints. The RFID tag can also indicate or alert if there are too many items of a particular product line that have been returned due to quality issues or customers' lack of interest (Senthil Kumar & Kumar 2010), allowing the fashion brand to not purchase that product in the future.

7.3.14 *Traceable warranties*

For luxury items like handbags, warranties are a must-have of commitment and service to customers. Items that are tagged by the RFID system can used to inform owners of the expiration of the warranty period. Customers can also track their items' information, like the location of the manufacturer or the authorized repairer while it is in transit for repair (Michael & McCathie, 2005).

7.3.15 *Customer behavior and product assortment*

An RFID system also enables retailers to learn about their customers' behavior. RFID antenna are attached to clothing racks or shelves to track every movement of items and customers. Every single time a consumer removes an item from the rack, data are sent for analysis. If consumers put a T-shirt back on the rack after just five seconds, it is likely they don't like the material. If that T-shirt has disappeared after more than five minutes from a rack, it can be assumed that the consumer is trying it on (Loebbecke & Palmer & Huyskens 2006). From the study of consumer behavior, the merchandiser will have a bigger picture of fast-moving items, slow-moving items and bestsellers as a knowledge base to conduct product forecasting for next season.

7.3.16 *Counterfeit prevention*

There has been tremendous growth of counterfeit products in the fashion and textile industry. Technological advancements have made it easy to copy any brand's logo and reproduce the same. Many of the rich and high-quality competitors take advantage of the technology and manufacture counterfeit products to get increased sales and to bring the brand equity down (Tuyls and Batina, 2006). Hence, counterfeit products are sold at higher prices and fail to meet consumer requirements.

Brand logos and the information on the labels of the items tell about the brand. However, these methods are not foolproof in the age of advanced technology. Fashion brands are always looking for advanced technological options to prevent the counterfeiting of products. Among the various options, DNA or genetic markers and microtaggants are the most important choices (Kwok et al., 2010). With DNA markers, botanical plants' DNA are embedded in the finished goods. The plant DNA is specific, which can help to uniquely identify the authentic product from the counterfeit.

High-end luxury fashion goods such as clothing, handbags, wallets and shoes are easier to counterfeit compared with many other products. High-end luxury brand's products face huge revenue losses due to counterfeit products. Hence, they take the problem seriously and try their best to resolve the problem. For example, the fashion brand Gucci has adopted a number of steps to prevent counterfeiting. In its handbag line, Gucci uses a digital certificate card integrated with an RFID chip. Customers can check the authenticity of the products by using a mobile or social media device. The company can also check for the counterfeit product, as the product codes are only stored in the company's database.

With microtaggants, microscopic particles ranging in size from 0.2 mm to 20 mm, containing unique numeric code sequences, are used to prepare the tag. They are often produced in multiple colors from plastic resins and adhesives. Generally, the producer of these microscopic particles is only known to a unique manufacturer, which makes it difficult to duplicate. The codes can be reproduced only after getting the permission of the brand owner. Among these two approaches, the DNA approach is more advanced and harder to copy.

In addition to the above two technologies, RFID technology can also be used to avoid counterfeit products. The data or logo relating to the brand can be stored in the RFID chip, which can be read by the reader from the radio frequency (Corbellini et al., 2006). An internal factory-programmed ID that can't be duplicated is used to store the information to avoid the counterfeit. The electronic signature of the brand can be embedded into two mathematically related keys: a secret key kept by the brand and a public key distributed to consumers. The private key is used for encoding the document and the public key is used for decrypting. When needed, the signature of the public key can be verified with the private key.

Counterfeiting is a sensitive problem to most fashion brands, especially luxury brands like Hermes and Chanel. Luxury fashion retailers work closely with law enforcement agencies to prevent "fake" retailers and manufacturers from mimicking their original products. However, they are still unable to completely stop the counterfeit products and need another solution to secure the trust of consumers. The use of RFID tags can help to reduce and prevent counterfeiting. When an RFID tag is attached to a product with a specific identification code, consumers can scan and authenticate their goods through smartphones or a secure website. This enables consumers to determine if their fashion item is fake or not. Furthermore, the use of RFID tags can help in managing the return of counterfeit products by consumers.

7.3.17 Fashion show

RFID wristbands are one of the latest innovations of RFID system application. It has been used at many concerts to prevent the use of fraudulent or counterfeit tickets. RFID technology is now starting to be used at fashion shows to improve security. Each band is attached with a unique tag that identifies each attendee's identity. This lets the event managers have full access and control and insight in real time about who is attending (Stewart and Paley, 2015).

7.3.18 Decisions on markdown

Markdown is the discount given by retailers to promote sales of some fashion goods that are not selling fast and stay for a longer period. Markdowns allow retailers to adjust prices in response to competition, overstocking, seasonality, shop-worn merchandise and decreased shop traffic. Although markdown represents a loss of profitability, the inventory is at least converted to cash, which is available as Open to Buy (OTB). Evaluating the causes of markdown is essential for planning and future buying. RFID technology can help in easily identifying the fashion goods that are selling fast and those are not selling or selling very slow. Once the peak period for selling a particular style has slowed down, merchandisers can decide on Monday morning meetings to provide discounts on those products. The information from RFID tags can help to make quick and accurate decisions on the right time for markdown in right quantity.

7.4 Opportunities and threats of RFID in retail

There are many opportunities as well as threats related to the use of RFID technology in retail. The use of RFID technology is rapidly increasing in many retail sectors. This will help in reducing the cost of installing RFID systems and getting services faster from RFID providers. The current use of RFID in retail is just 8% globally, which is very low. Many retail stores are still following traditional systems that were followed two decades ago. These retail stores receive the goods, physically verify the quantity and quality, allocate the stock keeping unit (SKU) with a barcode, while dealing with the products. This process is too time-consuming and error prone, which leads to revenue loss and reduced profitability.

Since the use of RFID technology is on the rise, retail stores not using RFID technology will see the benefits others are getting from it. They will gradually start adopting the technology. Hence, in future RFID technology will grow rapidly. This will help consumers receive products faster, and retail stores will benefit from the accuracy of available information and thus higher profitability.

There are no existing technologies that are currently a potential threat to RFID technology in the near future. Research and development may eventually bring about newer technologies that might ultimately prove to be superior to RFID technology. It is obvious that for the coming two to three decades, RFID technology has a bright future in retail sector.

Due to the use of RFID technology, retail stores achieve several benefits. However, there are some potential threats related to the use of RFID. With an RFID system, staff have less control over inventory. They lack up-to-date information on the physical location of fashion goods, their minimum stock level, replenishment level and replenishment frequency. This can lead to serious concerns such as loss of sales of some styles due to out-of-stock or excessive inventory for some styles.

In the out-of-stock situation, fashion goods might be located somewhere else from their designated shelf. Hence, when customers find the empty shelves, they drop the idea of buying the product. Out-of-stock situations can be avoided by retail staff checking empty shelves manually on a regular basis in addition to the data received from the RFID system. Being out-of-stock leads to a drop in net sales and thus net profits, in addition to lost customer loyalty. Similarly, excessive inventory can be avoided by manually checking the sales of each style on a regular basis and updating the stock of slow-moving styles. Excessive stock reduces overall stock turnover, as the cost of average inventory increases.

7.5 Conclusion

RFID technology is increasingly used in modern fashion retail stores. The use of RFID in fashion retail can help to easily manage several important jobs in retail stores, which includes product tracking, retail management, merchandise replenishment, reduction of shrinkage, prevention of theft, improved customer relationship management and counterfeit prevention. This can help to improve the efficiency and effectiveness of managing retail stores. The multifaceted benefits are realized through increased sales, better stock replenishment, reduced cost, reduced time, increased customer traffic and finally increased profitability.

There are many drawbacks in the use of RFID technology in retail, which creates hindrances in the implementation of RFID technology. The major drawback includes the cost of the RFID tags. In some instances, the cost of the RFID tags is even higher than the cost of the garment itself. Other problems are security and privacy issues, lack of standardization and security attacks. These technical problems can be resolved by research and development and developing standardardized rules. Furthermore, the increasingly use of RFID tags can help to bring the cost of the tags to an affordable range.

Many of the leading global fashion retailers have already implemented RFID technology in their retail stores and are realizing the benefits of

RFID technology. Several others are on the brink of adopting the technology after witnessing the benefits that others are enjoying. The future of RFID technology is bright in the retail sector. As the cost is reduced and newer application areas emerge, RFID technology will be a revolutionary technology in all the retail stores.

References

Age, C. S. 2005. Research Confirms Wal-Mart's RFID Benefit. *Chain Store Age*, 81, 80.

Azevedo, S. G. & Ferreira, J. 2009. RFID technology in retailing: an exploratory study on fashion apparels. *IUP Journal of Managerial Economics*, 7, 7-22.

Barnes, L. & Lea-Greenwood, G. 2010. Fast fashion in the retail store environment. *International Journal of Retail & Distribution Management*, 38, 760-772.

Battaglia, M., Testa, F., Bianchi, L., Iraldo, F. & Frey, M. 2014. Corporate social responsibility and competitiveness within SMEs of the fashion industry: Evidence from Italy and France. *Sustainability*, 6, 872-893.

Bhardwaj, V. & Fairhurst, A. 2010. Fast fashion: response to changes in the fashion industry. *The International Review of Retail, Distribution and Consumer Research*, 20, 165-173.

Brusey, J. & McFarlane, D. C. 2009. Effective RFID-based object tracking for manufacturing. *International Journal of Computer Integrated Manufacturing*, 22, 638-647.

Chen, J. V. & Pfleuger Jr., P. 2008. RFID in retail: a framework for examining consumers' ethical perceptions. *International Journal of Mobile Communications*, 6, 53-66.

Condea, C., Thiesse, F. & Fleisch, E. 2012. RFID-enabled shelf replenishment with backroom monitoring in retail stores. *Decision Support Systems*, 52, 839-849.

Corbellini, S., Ferraris, F. & Parvis, M. A cryptographic system for brand authentication and material traceability in the textile industry. Instrumentation and Measurement Technology Conference, 2006. Proceedings of the IEEE, 1331-1335.

Dai, H. & Tseng, M. M. 2012. The impacts of RFID implementation on reducing inventory inaccuracy in a multi-stage supply chain. *International Journal of Production Economics*, 139, 634-641.

De Kok, A., Van Donselaar, K. & Van Woensel, T. 2008. A break-even analysis of RFID technology for inventory sensitive to shrinkage. *International Journal of Production Economics*, 112, 521-531.

Gaukler, G. M., Seifert, R. W. & Hausman, W. H. 2007. Item-level RFID in the retail supply chain. *Production and Operations Management*, 16, 65-76.

Huber, N. & Michael, K. Minimizing product shrinkage across the supply chain using radio frequency identification: A case study on a major Australian retailer. International Conference on the Management of Mobile Business, 2007. ICMB 2007. IEEE, 45-45.

Jaselskis, E. J., Anderson, M. R., Jahren, C. T., Rodriguez, Y. & Njos, S. 1995. Radio frequency identification applications in construction industry. *Journal of Construction Engineering and Management*, 121, 189-196.

Jones, P., Clarke-Hill, C., Hillier, D. & Comfort, D. 2005. The benefits, challenges and impacts of radio frequency identification technology (RFID) for retailers in the UK. *Marketing Intelligence & Planning*, 23, 395-402.

Jones, P., Clarke-Hill, C., Shears, P., Comfort, D. & Hillier, D. 2004. Radio frequency identification in the UK: opportunities and challenges. *International Journal of Retail & Distribution Management*, 32, 164-171.
Koh, C. E., Kim, H. J. & Kim, E. Y. 2006. The impact of RFID in retail industry: issues and critical success factors. *Journal of Shopping Center Research*, 13, 101-117.
Kwok, S., Ting, S., Tsang, A. H. & Cheung, C. 2010. A counterfeit network analyzer based on RFID and EPC. *Industrial Management & Data Systems*, 110, 1018-1037.
Loebbecke, C. & Huyskens, C. 2008. A competitive perspective on standard-making: Kaufhof's rfid project in fashion retailing. *Electronic Markets*, 18, 30-38.
Loebbecke, C., Palmer, J. & Huyskens, C. 2006. RFID's potential in the fashion industry: a case analysis. *BLED 2006 Proceedings*, 34.
Medeiros, C. R., Costa, J. R. & Fernandes, C. A. 2008. RFID smart shelf with confined detection volume at UHF. *IEEE Antennas and Wireless Propagation Letters*, 7, 773-776.
Melià-Seguí, J., Pous, R., Carreras, A., Morenza-Cinos, M., Parada, R., Liaghat, Z. & De Porrata-Doria, R. Enhancing the shopping experience through RFID in an actual retail store. Proceedings of the 2013 ACM conference on Pervasive and ubiquitous computing adjunct publication, 2013. ACM, 1029-1036.
Michael, K. & McCathie, L. The pros and cons of RFID in supply chain management. International Conference on Mobile Business, 2005. ICMB 2005. IEEE, 623-629.
Moon, K. & Ngai, E. 2008. The adoption of RFID in fashion retailing: a business value-added framework. *Industrial Management & Data Systems*, 108, 596-612.
Nayak, R., Chatterjee, K., Khurana, G. & Khandual, A. 2007. RFID: Tagging the new era. *Man-Made Textiles in India*, 50, 174-177.
Nayak, R. & Padhye, R. 2014a. The care of apparel products. *In:* Sinclair, R. (ed.) *Textiles and Fashion: Materials, Design and Technology*. United Kingdom: Elsevier.
Nayak, R. & Padhye, R. 2014b. Introduction: the apparel industry. *In:* Nayak, R. & Padhye, R. (eds.) *Garment Manufacturing Technology*. Elsevier.
Nayak, R. & Padhye, R. 2015a. Care labeling of clothing *In:* Nayak, R. & Padhye, R. (eds.) *Garment Manufacturing Technology*. Elsevier.
Nayak, R. & Padhye, R. 2015b. *Garment Manufacturing Technology*, Cambridge, UK, Woodhead Publishing.
Nayak, R., Singh, A., Padhye, R. & Wang, L. 2015. RFID in textile and clothing manufacturing: technology and challenges. *Fashion and Textiles*, 2, 1-16
Reinhart, G., Philipp, T. R., Bernhard, J., Dräger, T., Grabowski, C. & Hoffmann, M. Integrating RFID Transponders into Fibre-Reinforced Plastics by Using Textile Production Technology. Smart Objects, Systems and Technologies (SmartSysTech), *Proceedings of 2012 European Conference on, 2012*. VDE, 1-6.
Rekik, Y., Sahin, E. & Dallery, Y. 2008. Analysis of the impact of the RFID technology on reducing product misplacement errors at retail stores. *International Journal of Production Economics*, 112, 264-278.
Rekik, Y., Sahin, E. & Dallery, Y. 2009. Inventory inaccuracy in retail stores due to theft: An analysis of the benefits of RFID. *International Journal of Production Economics*, 118, 189-198.
Roberti, M. 2006. RFID is fit to track clothes. *Chain Store Age*, 82, 158.
Roussos, G. 2006. Enabling RFID in retail. *Computer*, 39, 25-30.

Shin, S. & Eksioglu, B. 2015. An empirical study of RFID productivity in the US retail supply chain. *International Journal of Production Economics*, 163, 89-96.

Stewart, R. G. & Paley, D. N. 2015. Security system and method. U.S. Patent No. 9,071,447. 30 Jun. 2015.

Stratton, R. & Warburton, R. D. 2003. The strategic integration of agile and lean supply. *International Journal of production economics*, 85, 183-198.

Swedberg, C. 2013. *IER's expedited self payment kiosk speeds up checkouts* [Online]. Available: http://www.rfidjournal.com/articles/view?10667#back-from-modal [Accessed August 18 2018].

Tuyls, P. & Batina, L. RFID-tags for anti-counterfeiting. Cryptographers' Track at the RSA Conference, 2006. Springer, 115-131.

Twist, D. C. 2005. The impact of radio frequency identification on supply chain facilities. *Journal of Facilities Management*, 3, 226-239.

Uhrich, F., Sandner, U. & Resatsch, F. 2008. RFID in retailing and customer relationship management. *Communications of the AIS (CAIS)*, 23, 219-234.

Vowels, S. A. 2006. A straegic case for RFID: An examination of Wal-mart and its supply chain. In Proceedings of Ninth Annual Conference of the Southern Association for Information Systems, 148-152. Hyatt Regency Jacksonville, Florida.

chapter eight

Case studies of RFID adoption by famous fashion brands

KEYWORDS: *RFID technology, fashion brands, fast fashion, fashion supply chain, fashion retail*

8.1 Introduction

Generally, fashion brands are well known for being slow in adopting new technologies (Nayak and Padhye, 2014; Nayak and Padhye, 2017). This can be attributed to low product costs, the availability of cheap labor and the high initial investment costs for Radio Frequency Identification (RFID) technology (Nayak et al., 2015). However, many fashion brands have seen the real benefits of adopting this technology by other retailers such as Walmart and Tesco. Therefore, they want to embrace this technology into their supply chain to achieve the benefits of time and money saving. Among the applications of RFID technology as mentioned in Chapter 1, the major attention of fashion brands is centered at: resolving counterfeit products, improving customer experience, tracking and tracking and improving logistics, retail operation, inventory management and event organizing.

Many fashion brands employ RFID technology to prevent or reduce counterfeiting, which leads to losses in revenue. For example, fashion brands in Europe lost about €26.3 billion due to counterfeit products (Ryan, 2017). Brands such as Moncler and Salvatore Ferragamo employ the RFID technology so that customers can check the RFID tags for original products. Other important aspects of using RFID technology are focused on customer service. Brands such as Rebecca Minkoff, Ralph Lauren, Adidas, Nike, Myer and Burberry employ RFID technology so that customers can get information on product history, which products suits them, assistance in the fitting room to select the right merchandise and self-checkout.

The previous chapters have discussed the history, technology and application of RFID technology and its benefits in the fashion supply chain. This chapter will focus on the use of RFID technology by many leading fashion brands. Different uses of RFID technology by global fashion brands and the benefits achieved by each brand are discussed in this chapter. The information for this chapter has been collected from fashion brands' websites, standard databases, trade journals and fashion journals.

8.2 RFID: Examples from fashion industry

8.2.1 General situation

The benefits of barcode technology in retail sector have been long proven by their return on investment (ROI). Now is the time to realize the benefits of the new technology in place of the superseded barcode technology. For apparel retailers, where thousands to millions of stock keeping units (SKUs) need to be managed annually, RFID is the perfect choice for fashion retailers and brands. Not only fast fashion brands, but also luxury fashion brands are focusing on this technology to prevent fakes which have cost about 10 percent of sales every year.

UTS Global (UTS-Global, 2017), which has been analyzing and observing the performance of RFID in the fashion industry, claimed that 2017 was the "tipping point" for applying RFID to the industry. They also predict that RFID will help retailers surpass other competitors who do not use RFID, with as much as four times more success. High-end luxury fashion products such as Gucci perfumes, Louis Vuitton bags and Burberry coats are the target of shoplifters; these items can be sold on the black market for high prices. Similarly, fake versions of these products can be duplicated by manufacturers. The use of RFID technology can save millions of dollars by preventing theft and counterfeit products (UTS-Global, 2017). UTS-Global further states that nearly 8 billion RFID tags will be needed for the global market in 2017, as shown in Figure 8.1.

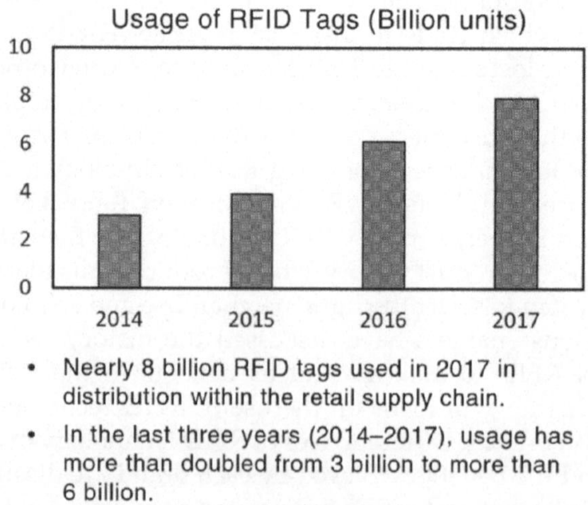

Figure 8.1 Number of RFID tags in apparel industry from 2014–2017.

8.2.2 Impact of RFID

The application of RFID technology in the supply chain has helped fashion brands in locating counterfeit products, improving supply chain efficiency, better control of inventory, shelf-replenishment of stock, increased sales, improved consumer service and reduced operational costs (Nayak et al., 2007a; Nayak et al., 2007b). Saarijärvi et al. (2013) states that there were 5% increase in the sales for fashion retailer after adopting the RFID technology. There is also a good performance of managing the inventory. Saarijärvi (2013) also states that specific and detailed information about size, color, price, code and other data related to a specific product can be held in an RFID tag, which makes it different from a barcode.

Another advantage of using RFID in fashion is that a product's information can be added anytime along its journey. Lalla (CEO of Nordic ID, Finnish handheld RFID computer manufacturer) (n.d., cited in Saarijärvi, 2013) states that RFID is also quick to install and has nearly 100 percent accuracy. Therefore, Isla (Inditex president) (Bendavid, 2015) claims that Inditex stores' operation has changed since using RFID. Figure 8.2 describes how the RFID technology can be implemented in a retail environment.

The benefits of implementing RFID technology in retail by leading fashion apparel retailers in the US and Europe are evident from the UTS-Global report (UTS-Global, 2017). As shown in Figure 8.3, it can be observed that inventory accuracy with RFID technology improved by 25% and out-of-stock situations were improved by about 41% due to

Figure 8.2 Usage of RFID in a retail store.

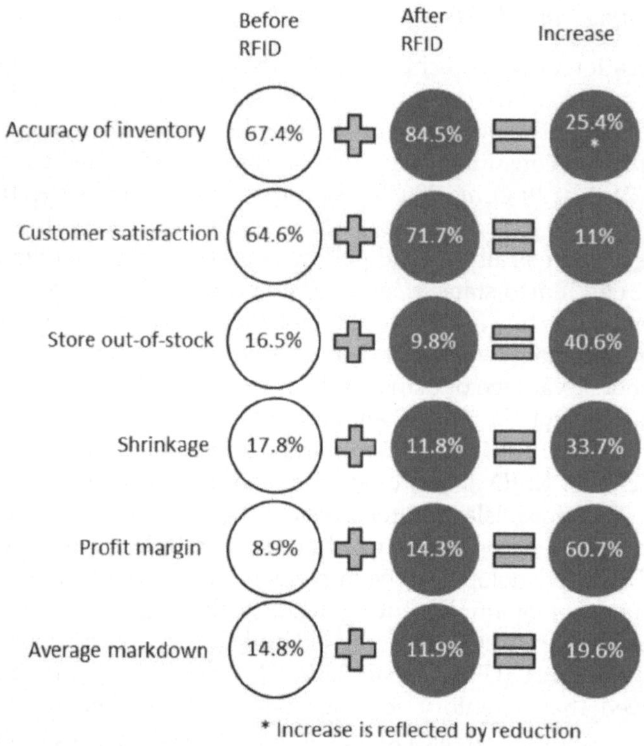

Figure 8.3 RFID impact in Retail KPI (key performance indicator).

quick stock replenishment, which impacted the key performance indicator (KPI) of the retail stores. Similarly, shrinkage was reduced by 34% and profit margins increased by 61%.

Overall, there are impressive benefits achieved by fashion retailers that use RFID technology, including improvement of inventory accuracy to customer satisfaction as well as profit margins. The benefits of RFID technology include:

- Reduced out-of-stocks by 50 percent.
- Improved customer experience and faster service.
- Provides valuable information about the supply chain.
- Inventory accuracy as well as increased speed of inventory management.
- Enables production monitoring and planning.
- Higher shipping accuracy includes reliable and fast receipt and shipping of goods.
- Minimizes error in sorting, creating a faster process.
- Reduces need for restocking, reducing labor costs.

- 90 percent reduction in labor associated with inventory counts.
- Reduction in receipt of counterfeit items.
- Reduced employee theft and shoplifting through improved monitoring of inventory movement.
- Faster receiving and inbound processing.
- Lost sales due to the inability to locate inventory are eliminated.
- Since each product in a specific shipment has a unique identifier, quality assurance and counterfeit prevention are enhanced.
- Improved accuracy in picking and packing operations translates into the proper fulfillment of purchase orders.
- Better control over own supply chain management.
- Faster billing process.
- Easy to create advanced shipping notice (ASN).

8.3 *Fashion companies that use RFID*

Many global fast fashion brands such as Zara (Inditex), H&M and Topman employ RFID in their supply chain activities. Similarly, traditional fashion brands such as Adidas, Nike, Thomas Pink, Macy's and Burberry are also adopting the technology in their supply chain, which is discussed in the following section.

8.3.1 *Inditex group & Zara*

Based in Spain, Inditex is a global fashion leader with more than 6300 stores worldwide. Inditex brands include: Zara (its biggest division), Bershka, Massimo Dutti, Pull and Bear and Stradivarius. The Inditex group owns all its design facilities in Europe and manufactures of almost all its clothing are based in Europe and other developing countries. At its annual meeting in 2014, the chairman and CEO mentioned the retailer's future RFID expansion plans for adopting the technology (Prince, 2014). Its major focus has been on store-to-store supply chain management involving all the upstream and downstream players. Since then, many of Inditex's brands (i.e., Zara, Pull and Bear and Massimo Dutti) have started using RFID technology. Item-level tracking is not the only purpose of Inditex while using RFID; other purposes include managing inventory, providing enhanced costumer experience and preventing retail shrinkage.

As of 2014, the chairman and CEO of Inditex stated that 700 stores of this brand used RFID within 22 countries. This fast fashion brand is aiming to implement RFID technology in all its supply chain activities by 2020 as outlined by its CEO (Barrie, 2018). Zara has spent about 1.5 billion Euros on technology upgrades such as RFID and logistics. The plan is to roll out the traditional method in phases and replace it with RFID technology.

Figure 8.4 Application of RFID Technology in Zara's supply chain.

The application of RFID technology by Zara has made the distribution process more efficient, provided more accurate results of retail inventory and improved customer service (Prince, 2014). Various applications of RFID technology in Zara's supply chain are illustrated in Figure 8.4.

Zara enjoyed a number of advantages after it introduced RFID into its supply chain. To be specific, every time an item was sold in one of its stores, the information from the chip is sent to the stockroom to update inventory levels. This helps in getting a real picture of the stock flow from the factory to store (Al-Kassab et al., 2013; Sanders, 2016). When stock arrived at Zara stores twice a week, the staff already knew which items needed to be restocked to the shelf. Previously, staff would have to replenish the shelf multiple times for one day followed by a hand-written report. With RFID technology, staff can easily assist customers to find the right product and right size or color according to the information from the RFID chips. They can access the system to determine whether the required products are still available in inventory or at other stores, resulting in better customer service and enhanced customer satisfaction (Bjork, 2014). This implementation works best for Zara since it applies vertical integration in the supply chain process (Sanders, 2016). Before introducing RFID into the supply chain and in its store, the inventory manager and a team of 40 had to check and barcode scan one product at a time. The total task took around five hours to complete (Bjork, 2014).

8.3.2 H&M

Recently the Swedish fashion retailer Hennes & Mauritz (H&M), one of the largest global fast fashion brands faced the problem of poor inventory management (Lopez, 2018) due to bad product cycle. This was attributed to H&M's poor digital capability for the last three years. They needed to adopt a better model suited to a digital environment for better inventory management. To resolve this problem, H&M is trying to adjust to the new market dynamics starting from 2018. In addition to strengthening their supply chain network, they are planning to install RFID technology in 1800 stores in 2019 (Lopez, 2018). The implementation of RFID technology can help automate warehouse management and improve product visibility and demand predictability.

H&M is also trying to adopt the supply chain principles of Zara to become successful in this business (McKevitt, 2017). H&M anticipates that the automation of warehouse management using RFID technology can help to resolve shortages and over-stocking issues. H&M is also planning to use RFID clothes tagging so that it is easier to manage the movement of the clothes (Ringstrom, 2017). This will also reduce the time consumed to track the products.

Recently, the renewal of RFID agreement by H&M with Nedap, shows H&M's effort for rapid growth of the technology to manage the inventory effectively (Dirks, 2017). Nedap will focus on intelligence item-level surveillance, installations of RFID in the new stores and refurbishment. This will help H&M to effectively manage inventory, prevent loss and provide on-time information to the brand and the central office.

8.3.3 Moncler

The French Luxury fashion brand Moncler started to use RFID for their Spring/Summer 2016 collection (Donaldson, 2016). This is primarily focused on the prevention of counterfeit products. The RFID tag of a garment holds individual information, which helps customers to check the authenticity of the products by scanning them with smartphones or at Moncler's website (or code.moncler.com). With the same technology that is involved with mobile pay apps, users of Apple Pay can identify whether the Moncler product they have purchased is a fake or not. The RFID technology uses a QR code in addition to the near field communication (NFC), using tags like Moncler's logo, which is shown in Figure 8.5.

Figure 8.5 Moncler's RFID tag resembles Moncler's logo.

Source: Moncler-Expert, 2018.

8.3.4 Adidas

Adidas started using RFID technology in Europe in 2013 in its NEO stores in Nuremberg, Frankfurt and Helsinki. The objective was to provide information to the customers in changing rooms, automatic payment at the point-of-sales (POS) and security (to prevent theft) in retail and inventory management. In 2016, Adidas installed the RFID system in 450 stores in Russia to increase its basic stock accuracy, on the floor availability and improved customer service (Binns, 2017). Adidas was facing the problem of poor inventory management, narrow range of products and poor customer service. The installation of RFID can improve the speed of service and improve customer satisfaction. Adidas also started using the RFID in its products such as soccer balls and sports shoes.

8.3.5 Burberry

RFID was introduced by Burberry in its retail stores in 2012 (Davis, 2014). This technology has been implemented in a few of its flagship stores. The major objective was stock management and quality control by using the RFID tags, which helps to provide an overview of store situation. Moreover, customer experience is also enhanced while using this technology by turning on the mirrors in the changing room when there are customers trying a product. A video is shown on the mirrors about how the items are made, as well as catwalk footage of the items. The tags are deactivated when the purchase is made by the consumers. However, these tags can still be present on the products if customers are happy in order to track the product life cycle and information on the usage pattern of the product. When the customer makes the purchases, the RFID tags are deactivated and reused in other garments. A virtual fitting room with RFID technology used by Burberry has been shown in Figure 8.6 (Wikipedia, 2013).

8.3.6 Nike

Like Adidas, Nike has started using RFID technology in its sports shoes (Menezes, 2008). Nike is leading in Canada in RFID applications and trying to implement the technology in India and China. The implementation of RFID in running shoes can provide detailed information on the events these shoes are being used with the assistance from the user. RFID tags used in shoes can give information on the time taken to complete a marathon. Nike sees great potential for RFID technology in its sportswear market.

Figure 8.6 Burberry's virtual fitting room with RFID technology.

Source: Intel Free Press, 2013.

8.3.7 Thomas Pink

Thomas Pink, the British luxury apparel is using the internet of things (IoT) based on RFID (UHF) technology for tracking people and product movement within brand's New York City stores, which is the greatly helpful to the brand (Swedberg, 2017). For example, a product is picked up by a customer and is carried to a separate location on the store. The customer then leaves this product in that location and selects another similar product. A second customer is looking for this abandoned item and cannot find it. A sales associate will be the one searching for this wanted product. RFID helps the associate to locate the exact place of this missing item, improving inventory management and customer service. Stores can react quickly, whether a product purchased is the last one of its style or size, RFID can alert this situation before it happens (UTS Global, 2017).

8.3.8 Macy's

Macy's aim is to track the location of products in its supply chain or in its stores, which lead to the use of RFID in 2011. Macy's senior VP mentioned that RFID technology will be implemented in all its company's stores by 2018 (Robert, 2016). Macy's also asked all its suppliers to adopt RFID technology with passive UHF tags. Since 2011, this technology has helped Macy's to fill online orders, which has helped in the accuracy of its order information and at a quicker rate. Moreover, it can track a specific item that is ordered on the brand's website for packing and shipping for customers to pick up at a store.

8.3.9 Myer

Myer is the largest department store in Australia for fashion and styling for the last 100 years. The giant retailer with 61 stores in Australia started to proof the concept in one of its stores (Elmas, 2017). In addition, Myer also extended its online, digital and mobile platforms to enhance customer experience. With 11 product categories and a growing market, it was hard for Myer to manage the business in 2016. Hence, the group started to test and implement the technology for inventory accuracy, reduced shrinkage and reduced labor costs. During their trial with an RFID system, significant benefits were achieved in these areas. Hence, the retail giant will implement the system in stores in strategic locations in Australia.

8.3.10 Costco

Costco, an American wholesale corporation, is the second largest retailer in the world after Walmart. While many fashion retailers have witnessed the benefits of using RFID technology, Costco thinks RFID can enhance the process, but can't replace the manual intervention (Roberti, 2013). They have found that the accuracy of data collected from the RFID technology can go up to 99%, while reading the items in the shopping cart. They are extremely good, but not good enough to replace cashiers. Costco does not suggest that RFID is not efficient. Their main emphasis is that human intervention is still needed for the RFID technology.

8.3.11 Ance Studio and RFID application

Ance Studio, a Swedish luxury fashion brand, has applied RFID to manage the inventory for its online retailing channel at total 45 stores internationally. There is difficulty in managing the stock level due to the fluctuation of sold items online and the inventory needed to be up to date at all times (Swedberg, 2017). The installation of RFID has enabled Ance Studio to maintain accurate stock levels as well as to visualize exactly which items are being sold on their online stores, improving its speed and sales performance. To be specific, the RFID system allows the company and staff to know which items are in the warehouse or in any specific store. Therefore, when there is an online purchase of a specific product, the order will be sent to the closest store that has the product then delivered to the customer (Swedberg, 2017). Furthermore, the brand also embraces the goal of "never out-of-stock" for customer benefit. To achieve this goal, the company had to hire more staff, with increased working hours at every store, for counting stock; however, this was not only financially inefficient and time consuming, but also prone to human errors. This situation improved after the installation of an RFID-based system, since it can place orders for replenishment when minimum inventory levels are reached (Swedberg, 2017).

8.3.12 *Lululemon*

In-store revenues for Lululemon, a global athletic apparel brand, have increased since implementing RFID technology (Swedberg, 2016). This deployment has been applied to all of its stores with the aim of being able to track products' information such as when they arrive at stores, when they are displayed on shelves and when they are sold. Aitken (RFID program director of Lululemon) (Swedberg, 2016) has said that Lululemon inventory accuracy has increased to 98 percent. The stores get the information of what goods need to be restocked on their shelves and are processed accordingly from the stockroom. This improvement in inventory accuracy has helped the brand to increase its ecommerce profits. Furthermore, the implementation of RFID technology has helped to save time in counting the inventory, increased merchandise visibility and helped customers to find the products they want. There have been 3.5 million products tagged at the brand's three distribution centers and their third-party warehouses; the suppliers have tagged around 6.7 million tags per annum.

8.3.13 *Salvatore Ferragamo*

Salvatore Ferragamo, the Italian fashion retailer, is using RFID technology to avoid counterfeit products being sold in the market (Taylor, 2016). The company started using RFID technology with women's shoes. They want the customers to buy the original product after checking the authenticity of the products from the RFID system. RFID-enabled microchips have been inserted in the products since 2014, which not only allows them to know the authenticity of the product, but also to track their movement. The customers can use a mobile app or web service to scan their purchased items and know whether they are original or not. The RFID technology has been extended to other products such as silk apparels, leather products, watches, fragrances and eyewear. By installing RFID technology, the brand was able to identify about 91,000 counterfeit products being sold online by different websites and seized 12,500 fake products from physical stores (Taylor, 2016). Chinese customs also intercepted 12,400 counterfeit products worth USD $17 million leaving the country.

8.4 Conclusion

RFID has been widely used by several global fashion retailers. Many of the fashion brands such as H&M, Zara, Adidas, Nike and Macy's have started using the RFID technology and felt the benefits. The application areas and benefits are numerous, which include: accuracy of inventory management, automatic stock update, increased customer service, prevention of theft and counterfeit products. However, some of the retail stores such as Costco feel that RFID technology can enhance the accuracy and save the time, but can't replace cashiers or that it can't work without human intervention. Some retail stores such as Myer have felt the benefits and started implementing the technology. In the future, many other retail stores will realize the benefits and adopt the technology.

References

Al-Kassab, J., Thiesse, F. & Buckel, T. 2013. RFID-enabled business process intelligence in retail stores: a case report. *Journal of Theoretical and Applied Electronic Commerce Research*, 8, 112-137.

Barrie, L. 2018. *Inditex to deploy RFID technology across all brands by 2020* [Online]. Available: https://www.just-style.com/news/inditex-to-deploy-rfid-technology-across-all-brands-by-2020_id134124.aspx [Accessed 17 July 2018].

Bendavid, Y. Assessing the Potential of RFID Technologies on the Productivity of a Freight Forwarder. *International Conference on E-Technologies*, 2015. Springer, 255-268.

Binns, J. 2017. *How Lululemon and Adidas use RFID to improve customer experience* [Online]. Available: https://risnews.com/how-lululemon-and-adidas-use-rfid-improve-customer-experience [Accessed 10 June 2018].

Bjork, C. 2014. *Zara builds its business around RFID.* [Online]. Available: https://www.wsj.com/articles/at-zara-fast-fashion-meets-smarter-inventory-1410884519 [Accessed 1 July 2018].

Davis, B. 2014. *Five retailers using NFC and RFID to enhance shopping: but do they work?* [Online]. Available: https://econsultancy.com/blog/65307-five-retailers-using-nfc-and-rfid-to-enhance-shopping-but-do-they-work [Accessed 11 August 2018].

Dirks, J. 2017. *H&M renews global multi-year contract with Nedap* [Online]. Available: http://www.nedap-retail.com/hm-renews-global-multi-year-contract-nedap/ [Accessed 1 July 2018].

Donaldson, T. 2016. *Moncler adds RFID to its products to fight counterfeiting* [Online]. Available: https://sourcingjournal.com/topics/technology/moncler-adds-rfid-to-its-products-to-fight-counterfeiting-td-44291/ [Accessed 11 July 2018].

Elmas, M. 2017. *Myer dips toe into RFID* [Online]. Available: https://www.insideretail.com.au/news/myer-dips-toe-into-rfid-201709 [Accessed 15 August 2018].

Lopez, E. 2018. *H&M bay cools its freight with RFID* [Online]. Available: http://www.rfidjournal.com/articles/view?4923 [Accessed 10 July 2018].

McKevitt, J. 2017. *H&M pursues fast fashion leader Zara through supply chain overhaul* [Online]. Available: https://www.supplychaindive.com/news/H-M-fast-fashion-zara-uniqlo-supply-chain-logistics/439410/ [Accessed 11 July 2018].

Menezes, J. P. 2008. *An RFID tag in your Nike shoes can win you that marathon* [Online]. Available: https://www.itbusiness.ca/news/an-rfid-tag-in-your-nike-shoes-can-win-you-that-marathon/1308 [Accessed 15 August 2018].

Moncler-Expert. 2018. *From the same jacket* [Online]. Available: http://www.monclerexpert.com/qrcode.html [Accessed 19 August 2018].

Nayak, R., Chatterjee, K., Khandual, A. & Gupta, T. 2007a. Scope & limitations of RFID. *Indian Textile Journal*, 117, 25-29.

Nayak, R., Chatterjee, K., Khurana, G. & Khandual, A. 2007b. RFID: Tagging the new era. *Man-Made Textiles in India*, 50, 174-177.

Nayak, R. & Padhye, R. 2014. Introduction: the apparel industry. *In*: Nayak, R. & Padhye, R. (eds.) *Garment Manufacturing Technology*. Elsevier.

Nayak, R. & Padhye, R. 2017. *Automation in Garment Manufacturing*, Woodhead Publishing.

Nayak, R., Singh, A., Padhye, R. & Wang, L. 2015. RFID in textile and clothing manufacturing: technology and challenges. *Fashion and Textiles*, 2, 1-16.

Prince, P. 2014. *Inditex CEO announces RFID expansion plans* [Online]. Available: http://www.rfidjournal.com/articles/view?11993 [Accessed 1 July 2018].

Ringstrom, A. 2017. *H&M invests in supply chain as fashion rivalry intensifies* [Online]. Available: https://www.reuters.com/article/us-h-m-results-idUSKBN1710PM [Accessed 8 August 2018].

Robert, M. 2016. *Macy's to RFID-tag 100 percent of items* [Online]. Available: http://www.rfidjournal.com/articles/view?15081/ [Accessed 21 May 2018].

Roberti, M. 2013. *Storefront Backtalk gets its RFID facts wrong—again* [Online]. Available: http://www.rfidjournal.com/blogs/rfid-journal/entry?10484. [Accessed 21 May 2018].

Ryan, R. 2017. *RFID Technology: 5 ways fashion brands are using it* [Online]. Available: https://promomagnews.com/blog/2017/10/13/rfid-technology-5-ways-fashion-brands-are-using-it. [Accessed 30 May 2018].

Saarijärvi, H., Karjaluoto, H. & Kuusela, H. 2013. Customer relationship management: the evolving role of customer data. *Marketing Intelligence & Planning*, 31, 584-600.

Sanders, N. R. 2016. How to use big data to drive your supply chain. *California Management Review*, 58, 26-48.

Swedberg, C. 2016. *RFID brings Lululemon's inventory accuracy to 98 percent* [Online]. Available: http://www.rfidjournal.com/articles/view?14354 [Accessed 13 May 2018].

Swedberg, C. 2017. *Thomas Pink piloting IoT system with RFID and cameras to manage inventory, traffic* [Online]. Available: http://www.rfidjournal.com/articles/view?15579 [Accessed 28 July 2018].

Taylor, P. 2016. *Ferragamo expands RFID tagging of products* [Online]. Available: https://www.securingindustry.com/clothing-and-accessories/ferragamo-expands-rfid-tagging-of-products/s107/a2714/#.W3NqRvZuLmI [Accessed 14 June 2018].

UTS-GLOBAL. 2017. *RFID success strategies for apparel retailers* [Online]. Available: https://www.ust-global.com/retail-rfid-success-strategies-apparel-retailers [Accessed 1 July 2018]. Wikipedia. 2013. *Virtual dressing room.* [Online]. Available: https://en.wikipedia.org/wiki/Virtual_dressing_room [Accessed 14 July 2018].

chapter nine

Challenges and future directions of RFID technology

KEYWORDS: *tag cost, standardization, compatibility, RFID technology, ethical issues*

9.1 Introduction

The global fashion and textile industries are becoming highly competitive. In order to gain competitive advantage, fashion and textile industries are now focused on the use of automation (Nayak and Padhye, 2017; Nayak and Padhye, 2018b), artificial intelligence (Nayak and Padhye, 2018a) and technology to improve the quality of garments. Radio Frequency Identification (RFID) technology is being widely used in many application areas in fashion and textile supply chain management to help the industry better manage its resources (Nayak et al., 2015). Many global fashion brands have already adopted the technology in their retail and manufacturing facilities. In future, other brands will adopt the technology depending on the RFID frequency of its operation, as described in Table 9.1.

In spite of the widespread applications described above, there are several limitations to this technology that need to be resolved before it can be more widely used in the industry, including: (Wu et al., 2006; Jones et al., 2005; Welbourne et al., 2007; Sheng et al., 2008):

- High investment cost
- Security and privacy issues
- RFID attacks by hackers
- Tag damage
- Fake tags
- Security attacks
- Compatibility
- Technological disadvantages
- Lack of standardization
- Lack of robustness
- Ethical issues
- Data management issues

Table 9.1 Forecast of RFID technology in the future

Low Frequency (LF) passive RFID system	Ultra High Frequency passive (UHF) RFID	Active RFID systems
Plant fiber farming (cotton, flax and hemp)	Fashion retail outlets	Library management
Animal fiber farming (wool and silk) and animal tracking	Animal tracking on farms, such as sheep and alpaca	Fashion and textile retail management
Tracking jail inmates, patients in hospitals and newborn babies	Managing assets in retail and manufacturing sectors	Warehouse management
Security access at entrances	Tracking people in hospitals and jails	Tracking people in hospitals and jails
	Fashion and textile logistics and warehouse management	Fashion supply chain and logistics
	Managing garments' transit in shipment and during storage	Managing garments' transit in shipment and during storage

9.2 Challenges to RFID

The following section describes the above-mentioned challenges or limitations of the RFID technology in detail.

9.2.1 High investment cost

One of the major challenges of implementing RFID technology is the cost of RFID tags, which is substantially higher than the barcode system (Ohkubo et al., 2005; Jones et al., 2005). The cost depends on the volume of usage: the higher the usage, the lower the cost will be. For example, the lowest price of commercially available RFID tags is 7.2 cents/tag, if the volume purchased is 10 million units or higher, which is rather a difficult task.

In general, the average cost of an RFID tag is around 50 cents if the volume is not sufficiently large. However, the cost of a barcode is just 5 cents, which many industries have set as their benchmark price for adopting the technology. When dealing with luxury fashion, an additional average cost of 50 cents does not make much of a difference in the product cost. However, for cheaper fashion products, which are sold in the range of $1–10, that 50 cents can cause the overall cost to increase substantially.

In addition to the cost of the tags, there are additional costs for implementing the reader, the software and the hardware to make the system operational (Wu et al., 2006). Therefore, retail industries are concerned about the impact on their return on investment (ROI) and net profits from implementing an expensive RFID system. Companies such as Walmart, Benetton and Prada are always focusing on reducing the price of tags as low as possible. If the price is reduced to an acceptable level, many fashion retailers and manufacturers can implement the technology. Furthermore, there is an additional cost involved in the maintenance and running of the system, which increases the overall cost (Nayak et al., 2007).

The use of RFID technology is worthwhile in retail sectors where the volume of sales is substantially higher. This can reduce queuing time for customers, prevent shrinkage and help in automatic replenishment of inventory. However, for stores with average or poor sales, RFID technology may not be the right choice due to cost factor. As many fashion brands are mandating RFID tags in their supply chain, all the downstream players need to embrace the technology in spite of the higher cost.

9.2.2 Security and privacy issues

Since the advent of RFID technology, issues of security and the use of unauthorized readers threatening consumers' privacy has been debated (Sarma et al., 2003, Peris-Lopez et al., 2006; Gao et al., 2004). There is a great challenge to managing the privacy of consumers in emerging countries. In developed countries such as the USA, a complex hierarchal structure is followed to protect consumer privacy. The wide use of RFID technology can negatively impact consumers' privacy.

When consumers purchase any product tagged with an RFID tag, their identity can be traced easily. During this process, the RFID tags use radio signals to broadcast the serial number or Electronic Product Code (EPC) to a nearby reader, which can be easily tracked by other readers programmed by hackers. The personal profile created by RFID technology can include names, personal choices, bank details and other personal information. Consumer profiles are created by fashion brands to take care of their target customers; they are not 100% safe with RFID technology. Hence, there is a high chance of consumer privacy violations.

In fear of consumer privacy breaches, several civil groups, such as the Electronic Privacy Information Center (EPIC) and the Electronic Frontier Foundation (EFF), are lobbying against the use of RFID technology. Due to this problem, the clothing retailer Benetton has planned to withdraw the use of RFID tags in its retail stores. CASPIAN (Consumers Against Supermarket Privacy Invasion and Numbering) group, based in the USA, has claimed that RFID tags embedded in items can track consumers, creating unwanted surveillance and an invasion of privacy. These issues need to be addressed and information needs to be provided to consumers while implementing the technology.

Many security measures have been proposed for RFID to overcome the above problem (Peris-Lopez et al., 2006; Osaka et al., 2006). The simplest approach is the "Kill Tag" approach, in which the tag is electronically deactivated when the item is sold, before the item is placed in the consumer's hand. In that scenario, the tag is dead and cannot be activated again. The tag can also be removed physically from the product before it is placed in the hand of the consumer. Another approach, which is easy to adopt, is the legislation approach, where strict restrictions are applied on the use of RFID technology. Legislation on the use of RFID tags will limit the collection of personal identifiable information and establish guidelines on the use of RFID technology.

Some other approaches to address security and privacy issue are: the cryptographic approach, the Hash Function approach, the Faraday Cage approach, the Randomized Hash Lock, the Backward channel XORing, the Active Jamming approach and the Standardization approach. It is essential for industries to adopt one or more of these approaches to address consumers' privacy concerns.

9.2.3 RFID attacks by hackers

RFID tags used in a range of applications are vulnerable to a range of security threats from hackers, which includes information tracking, cloning of tags and disabling tags. Sophisticated programs by hackers can be used to attack RFID systems.

9.2.4 Information tracking

Hackers can obtain product information by using a RFID reader similar to the one used in the store. The hackers can use RFID tag and frequency range similar to the tag that is in operation. This helps the hackers to design readers that can collect the radio signals and decode them. Although it is a difficult task, any shortfall in managing the back-end servers by the information technology (IT) team leads to this problem.

This type of attack can be prevented by the IT team by designing the tags in such a way that the tags respond in a different way to unauthorized readers. For example, the RFID tags can send unauthorized readers random information. Hence, the hackers may not be able to get the correct information.

9.2.5 Disabling the tags

Hackers can break the security algorithm used in the RFID system and disable RFID tags. Hence, the tags cannot respond to the signals from the reader and are permanently damaged. Since the tag signal can be accessed from several meters outside a store, hackers find it easy to operate away from the store and disable tags. Similarly, hackers can modify the information stored in the tags, which provides misleading information when the product is sold. These types of attacks are serious threats to modern RFID technology. The disabled RFID tag may not alert the storekeeper in case of a theft. The stores where RFID tags are used for automatic replenishment will no longer be trusted, as the true stock level won't be available.

These types of attacks can be prevented by using key cryptography for RFID tags. Each tag contacts the main server with a private non-erasable identifying key. When the attackers try to get the information from the RFID tags, a private key is needed. This technology cannot be used with cheaper RFID tags.

9.2.6 *Cloning of the tags*

Hackers can collect the information stored in RFID tags including the identification code of the tags. This helps the hackers to redesign the tags for use in counterfeit products. This is more common in luxury fashion products, where there is a high profit margin (Tuyls and Batina, 2006) and undermines fashion brands' efforts to prevent counterfeiting of their products. The cloning of RFID tags can be prevented with a similar approach to the use of key cryptography discussed above, where a specific key is needed for each tag to collect tag identification information.

9.2.7 *Side-channel attacks*

This type of attack happens during the implementation of the RFID system, where the hackers collect timing information and power consumption by using cryptanalysis. This type of attack can reveal the cryptographic keys to the hackers. This type of attack can be prevented by implementing secured RFID hardware network or secured cryptographic technology.

9.2.8 *Tag damage*

RFID tags can be physically damaged due to improper handling or by some perfidious customers. Some disloyal customers try to damage the tags. This type of damage can prevent data communication with the reader. RFID tags are also damaged due to electrostatic charge generation. When the RFID tags are attached to a garment with synthetic fibers or laminated to any adhesive web, the static charge can damage the tags. Eliminating the electrostatic charge can be the solution for this type of problem.

9.2.9 Compatibility

A lot of time and money is needed to set up RFID systems, which may not be economical for many companies. Furthermore, RFID may not be compatible with some existing technologies. Even if it can be integrated with existing systems, the expectation of 100% data accuracy is still a matter of concern (Smart et al., 2010; Lee and Chan, 2009). This becomes more important when the suppliers and buyers have to deal with their overseas counterparts. Some companies are afraid of using RFID technology, as the high investment may be replaced by other technologies as technological advancements in the near future may bring about better technologies than RFID.

9.2.10 Data management

The use of RFID technology involves managing a large amount of data relating to product information and business applications. To manage this data, a strong information technology team and the necessary software and hardware tools are needed. However, the majority of RFID adopters are only focused on the cost and storage capacity of the tags, neglecting the requirement of a data management team and devices. Without these there can be frequent failure of the RFID technology during the working period.

Organizations adopting the technology should emphasize the effective management of the spectrum of data generated by the system. Furthermore, organizations with multiple branches at different locations need to share data as soon as possible for updated information. The accurate sharing of data can create an additional burden for the IT team. Sharing of data with upstream players in the supply chain can lead to divulging important information and raises questions on the credibility of the data. Fashion retail stores with a large variety of products will use a wide range of RFID tags, creating additional work for the IT team.

9.3 Technological disadvantages

The application of RFID technology is still in its infancy. There has been much research, which is still ongoing, to improve the design and effectiveness of RFID tags. In addition, the software and hardware tools that integrate RFID technology are immature. More research is needed to overcome the disadvantages and gain wider acceptance of the technology.

Since RFID is based on the concept of radio waves, it can be interfered with by other radio signals (Bunduchi et al. 2011; Wu et al., 2006; Ni et al., 2011). For example, in a retail store, there may be several readers working at different frequencies. The signals from different readers can interfere with each other and provide wrong information. The degree of interference depends upon the frequency of the tag and the environment, which may lead to inaccurate reading. Hence, the use of RFID readers to read the price tags of purchased merchandise at the checkout counters may be interfered with by radio waves from other merchandise.

Furthermore, the RFID system can also be affected by the presence of metals and liquid. Low-frequency RFID tags can be obstructed by the presence of metal objects, and the high-frequency waves can be obstructed by the presence of liquids. On some occasions, inappropriate orientation of the tags in the merchandise cannot be detected by the readers. The tags, which are at a longer distance than the readable distance of the reader can not be read effectively. Hence, it can be a challenging task for the reader to collect the information. If several solid objects come across the signals sent to the readers from the RFID tags, the accuracy of the data will be impacted.

Other technological disadvantages are the scale of RFID operation, which is the capacity of an RFID reader to read multiple tags simultaneously. The limitations of the hardware, software tools and RFID interrogator to process a wide spectrum of data limit the scale of RFID operation. When several RFID tags need to be read simultaneously, interference between the RFID tag signals needs to be avoided. However, the anti-collision mechanism followed by many RFID systems is not very robust. It will take a huge amount of research and development to establish a robust anti-collision mechanism.

9.4 Lack of standardization

Barcode technology, which has been in operation for some decades, has already been standardized. However, there are many versions of RFID tags that operate at different frequencies and need different software, hardware and readers (Michael and McCathie, 2005; Goodrum et al., 2006; Bunduchi et al., 2011). The major factor is the frequency of the operation, which is mainly used on ultra-high frequencies (UHF). Due to this, RFID systems used in one country may not be functional in another. Differences between the existing standards create obstacles to standardisation, in addition to the reluctance of the RFID manufactures to adopt a single standard and share the royalties they earn. Therefore, RFID technology is in its infancy and there are many hurdles ahead of it. So, there is a need for a single agreed-upon frequency or group of frequencies to have interoperability between the manufacturers, retailers and distributors. Standardization of the RFID system will help in its wider acceptance.

Currently, two organizations, EPC Global and International Standards Organization (ISO) are working to standardize the use of RFID in the UHF spectrum (Wu et al., 2006). Although, standards such as EPC class 1

G2 (generation 2) protocol and ISO 18000-6 are being released, neither standard is yet finalised and there continues to be a lack of compatibility with each other (EPCglobal, 2004). EPC class 1 G2 is an international standard relating to RFID in the UHF range, which has been endorsed by the ISO. This royalty-free standard will help to develop a global supply chain network implementing RFID in the UHF range.

It has been publicly announced by EPC global that the EPC class 1 G2 standard has been endorsed by ISO in ISO/IEC 18000-6 amendment as Type C. Based on this standard, about 10 standardised tags and readers are commercially available. Therefore, a standardized system would provide necessary information and product to all RFID users in the supply chain, which will simplify the process.

However, more research and development are needed on resolving the technological issues and fixing the related standards. The effort of all standard organizations to develop and establish necessary standards related to their use, manufacture and technical specification can help to resolve the issues related to standardization. In addition, the technological developments and innovations in information technology will also assist in this direction to achieve wider adoption of RFID tags.

9.5 Ethical issues

It has already been discussed in Chapter 1 that RFID tags can be used for tracking people. This was approved by the food and drug administration (FDA) in 2004 in hospitals to identify patients with specific needs. The implantable VeriChip was approved to store personal information. Since then, RFID technology has been used for tracking patients, jail inmates and even newborn babies. The FDA approval to use implantable VeriChip has raised many questions on ethical and privacy issues. Some groups oppose their use on religious grounds, whereas others raise political or social grounds (Hunt et al., 2007; Wang and Loui, 2009). There have been concerns raised about personal identification security and medical data privacy maintained in the medical database. Before the wider acceptance of RFID technology, these ethical issues need to be addressed.

Due to all these technological disadvantages, RFID technology is considered immature. That immaturity can lead to inaccurate data delivery, which is a hindrance to the wider acceptance of the technology. It is certain that these technological disadvantages will be resolved one day, and RFID technology will work with full accuracy.

9.6 Research on RFID

As mentioned above, RFID technology has not matured. Software companies such as IBM, Oracle and Microsoft are spending a huge amount of money to achieve widespread application. Various research has been done on the technology including tags, antennae and readers of RFID tags (Nikitin et al., 2005; Ukkonen et al., 2012; Kellomäki et al., 2012; Occhiuzzi et al., 2010; Gimpel et al., 2004); communications infrastructure (Srivastava, 2010, Zuo, 2010, Welbourne et al., 2007); logistics (Wang and Wang, 2009, Kapoor et al., 2009); security (Fishkin et al., 2005, Simson et al., 2005, Thiesse, 2007; Yeo and Kim, 2005; Knospe and Pohl, 2004); and standardization (Moon and Ngai, 2008; Bolic et al., 2010).

Although there are several challenges at present to the adoption of RFID, they can be overcome with technical developments. Similar to other emerging technologies, RFID will become a reality in fashion and textiles over time. Research and development can help in mitigating the associated problems with the use of RFID. The technological issues such as 100% reliability can be achieved by: (1) installing multiple antennas in a tag with different orientations; (2) customizing the tag and antenna design for each material type; and (3) installing multiple readers to increase readability of the tags. Reading multiple tags simultaneously is possible using binary search techniques. Repetition of reading the tags can be avoided by putting the tags already read into sleep mode. Research and development will help in maximizing the antenna power pattern and the chances of successful reading (Weber, 2010).

Research is needed to explore how RFID technology can be used to improve inventory management performance by reducing forecast error and inventory discrepancy. Various models need to be developed to establish a suitable benchmark for calculating the value of the RFID system to improve inventory management (Burmester and De Medeiros, 2007; Nayak and Padhye, 2011). Research is needed to manage inventory operations in a supply chain where both the systems (bar coding and RFID) are working.

Furthermore, for successful application of RFID, existing research should focus on a host of disciplines, such as electronic engineering, information systems, computer science and business strategy (Bhattacharya et al., 2007). Before any research and development begins, it is vital to determine the principal concerns of RFID, whether technological, application related or security related. Maintaining the security of RFID tags is a challenging job. Improvements in information technology and the design of smart RFID tags can help to overcome this challenge.

9.7 Future trends of RFID in fashion supply chain

Retail giants such as Walmart, Tesco and Proctor & Gamble have already implemented RFID technology in their manufacturing and distribution networks. This technology has a bright future as it is being widely adopted by other retailers. Although the technology provides numerous benefits, the success of adopting the technology will mainly rely on the approach to resolving the challenges discussed in this chapter. These challenges may limit the realization of the full benefits offered by the technology.

Depending on the speed of implementing the technology, adopters can be classified into three groups: early adopters, fast movers and slow followers. Early adopters were much quicker in planning and implementing the technology than others. This group of retailers readily invested on the RFID technology at the early stage and was ready to accept the risks, if any. Many of the early adopters of RFID are realizing the benefits of implementing the technology. Fast movers started adopting the technology later, but quickly planned to implement the technology following the steps from the previous group. The slow followers have been quite slow and not ready to make a change in their management practices. They are not ready to make a huge investment considering the risk factors of the technology, if it becomes obsolete one day, in the future

The use of RFID systems in fashion and textiles is rapidly increasing, focusing on manufacturing, distribution and storage. Many fashion brands such as Tommy, Zara, Macy's and H&M have already implemented the technology. Other brands such as Mango and Dazzle fashion group are at the planning and implementing stage. In the future, an increased number of manufacturers, distributors and retailers in these businesses will be using an RFID system.

Although RFID systems cannot completely replace barcode technology, mainly due to higher cost, the accuracy, speed and return on investment of an RFID system will benefit retailers adopting the technology. RFID is a promising technology which will become ubiquitous in the future, helping organizations to solve their problems in supply chain management, security, personal identification, real time information availability and object tracking.

Success lies in understanding the technology and other features to minimize potential challenges. Direct benefits have been realized by companies such as GAP, the US apparel company, which increased its sales by 2% using RFID to reduce stock-outs. Another well-known US apparel company, J Crew, increased the speed of its inventory keeping time by five to eight times using RFID. Hence, it is high time that all fashion and textile industries start utilizing this new technology in various application areas such as manufacturing, supply chain, storage and retail and inventory management.

The apparel industry is characterized by short product life cycles, frequent style changes, a wide range of product designs, variable production volumes, high competitiveness and often high demand on product quality. In addition to these, the use of RFID increases cost and creates additional challenges for maintaining privacy of information. Organizations dealing with fashion and textiles should start by tracking expensive goods, as the decrease in labor costs and shrinkage can compensate the cost. RFID has advantages in speed, accuracy and convenience over traditional barcode systems, which are being realized by retail stores and manufacturers. In the future, RFID will gradually replace the traditional barcode systems in the fashion supply chain.

Important technical challenges discussed in this chapter still need to be resolved for successful operation of RFID in the fashion and textile sector (Prater et al., 2005). Furthermore, many industries are classified as small and medium enterprises for whom the high cost of implementing RFID is a major impediment. It will take some years or maybe a decade or so before the full benefits of RFID technology are realized. In future, we will increasingly see how RFID technology can extend the ability of computers in combination with the internet to sense and respond to the physical world. RFID technology will make computing an unobtrusive, intuitive part of everyday life, as refreshing as a walk in nature.

The largest challenge in textile and apparel manufacturing is the cost, which can be reduced by increasing the volume of usage. In addition, standardization and technological developments can help in reducing the cost. RFID chips can be developed that work in different places for different applications. These ideal RFID chips will require a frequency-independent capability the capability of being integrated with read/write memory. In addition, the software needs to be customized to accommodate the number of different uses. Currently, all merchandisers use barcodes for identification and at point-of-sale; the barcodes are easily printed and attached to the products within a very short period. Finalization of the design and preparation of RFID tags is a time-consuming process, which also needs to be carefully considered.

The issues related to standardization can be resolved by the effort of governing bodies to develop various standards relating to their use, manufacture and technical specification. Technological developments will also assist in this direction to achieve wide adoption of RFID tags. In addition, building an RFID-friendly infrastructure is a time-consuming process for many organizations due to the complexities involved with the process.

9.8 Conclusions

This chapter has highlighted the challenges and future directions of RFID applications in fashion and textiles. From the above discussions, it is evident that RFID technology suffers from several disadvantages and threats. The major disadvantage is the cost of RFID tags and the major threat is security attacks. For the wide acceptance of RFID technology in fashion retail and its supply chain, all the disadvantages and threats need to be resolved.

RFID technology has a very bright future in the fashion and textile supply chain. New areas of applications will emerge from technical advancements and we will see more RFID applications in departments where they are currently not being used. Fashion goods with high costs or with a high risk of being stolen will be the first to realize the benefits of RFID technology. In the future, the concept of "automatic checkout" may be a reality in many retail stores where the consumer's ID cards can just be scanned and payments will be managed through online payment gateways.

References

Bhattacharya, M., Chu, C.-H. & Mullen, T. RFID implementation in retail industry: Current status, issues, and challenges. Proceedings of 38th Annual Meeting of the Decision Sciences Institute, Phoenix, AZ, 2007. Citeseer, 2171-2176.

Bolic, M., Simplot-Ryl, D. & Stojmenovic, I. 2010. *RFID Systems: Research Trends and Challenges*, John Wiley & Sons.

Bunduchi, R., Weisshaar, C. & Smart, A. U. 2011. Mapping the benefits and costs associated with process innovation: The case of RFID adoption. *Technovation*, 31, 505-521.

Burmester, M. & De Medeiros, B. RFID security: attacks, countermeasures and challenges. The 5th RFID Academic Convocation, The RFID Journal Conference, 2007.

EPCGlobal 2004. Radio-Frequency Identity Protocols Class-1 Generation-2 UHF RFID Protocol for Communications at 860 MHz–960 MHz Version 1.1.0. [Online]. Available: https://www.gs1.org/sites/default/files/docs/epc/uhfc1g2_1_1_0-standard-20071017.pdf [Accessed 7 August 2018].

Fishkin, K. P., Roy, S. & Jiang, B. 2005. Some methods for privacy in RFID communication. *In European Workshop on Security in Ad-hoc and Sensor Networks*. Springer, Berlin, Heidelberg.

Gao, X., Xiang, Z., Wang, H., Shen, J., Huang, J. & Song, S. An approach to security and privacy of RFID system for supply chain. IEEE International Conference on E-Commerce Technology for Dynamic E-Business, 2004. IEEE, 164-168.

Gimpel, S., Mohring, U., Muller, H., Neudeck, A. & Scheibner, W. 2004. Textile-based electronic substrate technology. *Journal of Industrial Textiles*, 33, 179-189.

Goodrum, P. M., McLaren, M. A. & Durfee, A. 2006. The application of active radio frequency identification technology for tool tracking on construction job sites. *Automation in Construction*, 15, 292-302.

Hung, V. D., Puglia, A. & Puglia, M. 2007. *RFID: A Guide to Radio Frequency Identification*, John Wiley & Sons.

Jones, P., Clarke-Hill, C., Hillier, D. & Comfort, D. 2005. The benefits, challenges and impacts of radio frequency identification technology (RFID) for retailers in the UK. *Marketing Intelligence & Planning*, 23, 395-402.

Kapoor, G., Zhou, W. & Piramuthu, S. 2009. Challenges associated with RFID tag implementations in supply chains. *European Journal of Information Systems*, 18, 526-533.

Kellomaki, T., Virkki, J., Merilmapi, S. & Ukkonen, L. 2012. Towards washable wearable antennas: A comparison of coating materials for screen-printed textile-based UHF RFID tags. *International Journal of Antennas and Propagation*, 2012, 1-11.

Knospe, H. & Pohl, H. 2004. RFID security. *Information Security Technical Report*, 9, 39-50.

Lee, C. & Chan, T. 2009. Development of RFID-based reverse logistics system. *Expert Systems with Applications*, 36, 9299-9307.

Michael, K. & McCathie, L. The pros and cons of RFID in supply chain management. International Conference on Mobile Business, ICMB 2005. IEEE, 623-629.

Moon, K. & Ngai, E. 2008. The adoption of RFID in fashion retailing: A business value-added framework. *Industrial Management & Data Systems*, 108, 596-612.

Nayak, R., Chatterjee, K., Khurana, G. & Khandual, A. 2007. RFID: Tagging the new era. *Man-Made Textiles in India*, 50, 174-177.

Nayak, R. & Padhye, R. 2011. Application of modelling and simulation in smart and technical textiles. *In*: Patanaik, A. (eds.) *Modeling and Simulation in Fibrous Materials: Techniques and Applications.* Nova Science.

Nayak, R. & Padhye, R. 2017. *Automation in Garment Manufacturing*, Woodhead Publishing.

Nayak, R. & Padhye, R. 2018a. Artificial intelligence and its application in the apparel industry. In: Nayak, R. & Padhye, R. (eds.) *Automation in Garment Manufacturing.* Elsevier.

Nayak, R. & Padhye, R. 2018b. Introduction to automation in garment manufacturing. In: Nayak, R. & Padhye, R. (eds.) *Automation in Garment Manufacturing.* Elsevier.

Nayak, R., Singh, A., Padhye, R. & Wang, L. 2015. RFID in textile and clothing manufacturing: technology and challenges. *Fashion and Textiles*, 2, 1-16.

Ni, L. M., Zhang, D. & Souryal, M. R. 2011. RFID-based localization and tracking technologies. *Wireless Communications, IEEE*, 18, 45-51.

Nikitin, P. V., Lam, S. & Rao, K. Low cost silver ink RFID tag antennas. Antennas and Propagation Society International Symposium, 2005 IEEE, 353-356.

Occhiuzzi, C., Cippitelli, S. & Marrocco, G. 2010. Modeling, design and experimentation of wearable RFID sensor tag. *IEEE Transactions on Antennas and Propagation*, 58, 2490-2498.

Ohkubo, M., Suzuki, K. & Kinoshita, S. 2005. RFID privacy issues and technical challenges. *Communications of the ACM*, 48, 66-71.

Osaka, K., Takagi, T., Yamazaki, K. & Takahashi, O. An efficient and secure RFID security method with ownership transfer. International Conference on Computational Intelligence and Security, 2006. IEEE, 1090-1095.

Peris-Lopez, P., Hernandez-Castro, J. C., Estevez-Tapiador, J. M. & Ribagorda, A. RFID systems: A survey on security threats and proposed solutions. In: Cuenca P., Orozco-Barbosa L. (eds). Personal Wireless Communications, 2006. Springer, Berlin, Heidelberg.

Prater, E., Frazier, G. V. & Reyes, P. M. 2005. Future impacts of RFID on e-supply chains in grocery retailing. *Supply Chain Management: An International Journal*, 10, 134-142.

Sarma, S. E., Weis, S. A. & Engels, D. W. 2002. RFID systems and security and privacy implications. In: International Workshop on *Cryptographic Hardware and Embedded Systems-CHES 2002*. Springer, 454-469.

Sheng, Q. Z., Li, X. & Zeadally, S. 2008. Enabling next-generation RFID applications: Solutions and challenges. *Computer*, 41, 21-28.

Simson, L., Juels, A. & Pappu, R. RFID privacy: An overview of problems and proposed solutions. IEEE Symposium on Security & Privacy, 2005. 34-43.

Smart, A. U., Bunduchi, R. & Gerst, M. 2010. The costs of adoption of RFID technologies in supply networks. *International Journal of Operations & Production Management*, 30, 423-447.

Srivastava, B. 2010. Critical management issues for implementing RFID in supply chain management. *International Journal of Manufacturing Technology and Management*, 21, 289-307.

Thiesse, F. 2007. RFID, privacy and the perception of risk: A strategic framework. *The Journal of Strategic Information Systems*, 16, 214-232.

Tuyls, P. & Batina, L. RFID-tags for anti-counterfeiting. Cryptographers' Track at the RSA Conference, 2006. Springer, Berlin, Heidelberg, 115-131.

Ukkonen, L., Sydanheimo, L. & Rahmat-Samii, Y. Sewed textile RFID tag and sensor antennas for on-body use. 6th European Conference on Antennas and Propagation (EUCAP), 2012. IEEE, 3450-3454.

Wang, J. L. & Loui, M. C. Privacy and ethical issues in location-based tracking systems. IEEE International Symposium on Technology and Society, 2009. ISTAS'09. IEEE, 1-4.

Wang, L. & Wang, G. 2009. RFID-driven global supply chain and management. *International Journal of Computer Applications in Technology*, 35, 42-49.

Weber, R. H. 2010. Internet of Things–New security and privacy challenges. *Computer Law & Security Review*, 26, 23-30.

Welbourne, E., Balazinska, M., Borriello, G. & Brunette, W. Challenges for pervasive RFID-based infrastructures. Fifth Annual IEEE International Conference on Pervasive Computing and Communications Workshops, 2007. IEEE, 388-394.

Wu, N.-C., Nystrom, M., Lin, T.-R. & Yu, H.-C. 2006. Challenges to global RFID adoption. *Technovation*, 26, 1317-1323.

Yeo, S.-S. & Kim, S. K. 2005. Scalable and flexible privacy protection scheme for RFID systems. *Security and Privacy in Ad-hoc and Sensor Networks*. Springer, Berlin, Heidelberg. 153-163.

Zuo, Y. 2010. Survivable RFID systems: Issues, challenges, and techniques. *IEEE Transactions on Systems, Man, and Cybernetics, Part C: Applications and Reviews*, 40, 406-418.

Index

Note: *Italicized* page numbers refer to figures, **bold** page numbers refer to tables

M

N

O

P

Q

R

S